The Nebra Sky Disc

Cycles in the Cosmos

Howard Crowhurst

epistemea

Second edition 2023

Books in English by the same author published by Epistemea :

Carnac, The Alignments, When Art and Science were one, 2010.
Back to Square One, 2011
Carnac, 'twixt Heaven and Earth (photographs), 2017
The Megalithic Plan, 2021
52 Mysteries, The hidden science of the cards, 2023
Versailles, The Other Story, Vol.1 : The secret science of the Sun, 2023

Please visit howard.crowhurst.com for more information

Howard Crowhurst asserts his moral rights to de identified as the author of *The Nebra Star Disc, Cycles in the Cosmos* in accordance with the Copyright and Data Protection Act, 1988.

All rights reserved. No part of this book may be used or reproduced in any form or by any means, or stored in a database or retrieval system, without prior written permission of the publisher except in the case of brief quotations embodied in critical articles and reviews. Making copies of any part of this book for any purpose other than your own personal use is a violation of copyright laws.
This book is sold as is, without warranty of any kind, either express or implied, respecting the contents of this book, including but not limited to implied warranties for the book's quality, performance, merchantability, or fitness for any particular purpose. Neither Epistemea nor its dealers or distributors shall be liable to the purchaser or any other person or entity with respect to any liability, loss, or damage caused or alleged to be caused directly or indirectly by this book.

Proof reading : Marilyn Longstaff (many thanks)
Second edition, Copyright © EPISTEMEA, 2023

Copyright© for all diagrams, photos in figures 5 and 10, front cover (photos and design) : Howard Crowhurst, 2012
The photograph of the Nebra Sky disk reproduced here on the cover and inside is from Wikimedia commons and is accredited to Dbachmann with colour improvement by Rainer Zenz.
Photos ©: Figure 1: Chantal Jègues-Wolkiewiez, Figure 8: Brian Brondel, Figure 11: Google Earth

ISBN : 978-2-37900-033-1
EAN : 9782379000331
For any information or orders, please visit our website
www.epistemea.co.uk   or contact
EPISTEMEA, 4 avenue de l'océan, 56340 PLOUHARNEL, France.
contact@epistemea.fr 00 33 953 189 000

CONTENTS

PREFACE

The practice of prehistoric archaeology could be compared with doing several jigsaw puzzles simultaneously where the different pieces have been scattered in many countries over several centuries. When a piece of a puzzle is found by archaeologists during their excavations, they know more or less where to fit it, since the funding of any dig requires an a priori working hypothesis and they generally know what they are looking for. When the discovery comes as a total surprise, however, which is the case when non-specialists unearth something important by chance, it may put into question the current model. If it is realised that the piece will not fit correctly into any puzzle, then it may mean that a whole new puzzle comes into existence and that pieces which had been fitted, more or less comfortably, into one puzzle, may well fit much better in the new one.

Accepting the existence of some picture which has never been seen before is not an easy thing to do emotionally, especially as one gets older. Working on constantly moving foundations is incredibly difficult for all but sailors, who, from the beginning, are obliged to get over sea-sickness or change jobs. No matter how many stabilisers may be fitted, boats will always rock. All scientific disciplines are subject to this principle, but archaeology navigates in rougher seas than most and several well-known archaeologists have seen their established theories reduced to dust by some amazing discovery or the arrival of new scientific methods of analysis, such as carbon 14 or DNA.

The story of Ötzi, a man from -3300[1] who emerged from the ice in the Italian Alps in 1991, is a good example. Everything about him was surprising, even what he'd eaten for his last meal. He was carrying an axe head made from pure copper. This put the copper melting and shaping industry in Italy over 1000 years earlier than previously thought. His autosomal DNA seems to indicate that he came from the geographically isolated populations of Sardinia and Southern Corsica while his mitochondrial DNA could not be categorized and has given rise to a new subclade. In other terms, nobody alive today is a descendant of his genetic group. His clothes were much more sophisticated than had been previously imagined. His shoes were so well made that it is

---

1   All dates will be given in negative number of years before current era (BC). Positive numbers mean since the beginning of current era (AD)

believed that professional cobblers were already doing business. Twenty years after his frozen body was extracted from the ice, research is still revealing new and unexpected information.

In August 2007, Adam MacHale, a British tourist on holiday in Quiberon, near Carnac in Southern Brittany, found 4 large neolithique ritual axeheads in jadeite, dated from around -4500, as he was searching for shellfish on the shallow sea floor during an extremely low tide. It was the first discovery of its kind since the 19th century. The fact that these objects of great value (the jadeite comes from the Alps) had not been "buried" inside a tumulus leads to questions about their function. Despite their exposure to the tides "*their shape was so perfect I coudn't believe they were very old.*"[2] Adam MacHale actually cut himself on a blade of one of the axe heads as he was digging it out of the sea bed. This find then led to a new category of "submerged archaeological remains" in the Morbihan area of Brittany, which has brought to light several unknown underwater stone alignments and pushes back the dating of these megalithic monuments.

None of these discoveries enable us to know what was going on in these prehistoric peoples minds. If you put Einstein on a desert island and then study what remains from his stay 4000 years later, it is most improbable that you would find $E=MC^2$. Even if he engraved it on a stone which he buried in the ground, it wouldn't mean anything after four millenia. By studying Bouddha's clothes or the contents of his stomach, no indication of his inner state would be revealed. Were his DNA to be of a different nature, it would simply give rise to a new category, as for Ötzi.

At any given period, the current mindset is so overwhelming that it is virtually impossible to understand anything which could be radically different. This is obvious when we read what people thought about astronomy in the 14th century. It's much more difficult to see the problem for oneself in the present but it only takes a slight amount of humility to realise that the closed minds which were in the majority at Copernicus' time are most probably of a similar nature today, albeit on a different level of understanding.

What Copernicus proposed was a different way of looking at

---

2   Sonia Hoba, Adam MacHale's girlfriend. Ouest France, 18/09/2008.

things. Instead of positioning the Earth at the centre of celestial movement, he suggested a model where the Sun was immobile. As this made things a lot easier to understand, the model was eventually adopted, but only after quite a lot of pointless suffering. The heliocentric system ousted the geocentric view and became part of a radically new mindset. Unfortunately, this way of looking at things, which no intelligent person would dare question, can have rather perverse effects. Today, small children are taught to visualise the solar system from a point in space where they will never go. Their relationship with the heavens becomes one of mental representation instead of direct experience and consequently, fewer and fewer people actually know what is happening in their local night sky. Astronomy has become the business of incredibly complex and expensive equipment which enables one to plunge deeper and deeper into space, examined on screens. This is not without interest, but it is available only to an elite. "Common" folk have slowly but surely been cut off from what used to be an important part of the human realm.

In fact, the observer is always the centre of his own universe and as he looks up at the stars, he is confronted with the reality of his own existence. This perception of Me, here and now, incredibly small in an infinite universe, is a fundamental personal spiritual experience, common to humans throughout millennia and the essence of human consciousness. It explains in part why the Heavens have always played an important role in religious teachings and have been seen as the home of superior beings, the gods. This personal vision of the night sky, which one could call ego-centric, changes according to when and where the observation takes place. If the observer looks at the sky at different times of the year and from different latitudes, his own understanding may develop into a geocentric vision. This will allow him to deduce patterns, cycles and axes and realise that celestial objects obey a certain order. He will eventually become able to **foresee** events, a capability which can give a feeling of power and superiority, a totally different experience from that of pure consciousness.

The question of how this power may be used has modelled human society from the outset. The translation of Sumerian clay tablets has shown two major obsessions. The first is how to predict future events and the second is how to maintain secrecy. It is clear that the Sumerian leaders were very worried about their knowledge falling into the wrong hands. An incredibly long training programme with many hierarchical levels, destined to filter out unworthy candidates, awaited the future scribe in Egypt and Mesopotamia. The priesthood, the highest level, was all powerful and totally devoted to maintaining tradition.
These facts help us to understand the primordial role held by astronomy in the first human communities.

When students in archaeology start their first year, one of the first things they must establish is the difference between humans and animals. Once this list is established, students may choose to study certain branches of human activities that are of particular interest to them. Unfortunately, astronomy is not on the list in most European universities, although the human capacity to see the stars should be right at the top. Animals cannot see stars. Their field of vision does not take them in. The fact that archaeology students do not study elementary astronomy is a great handicap for them when they try to understand humankind's earliest representations and its architecture.
It is universally accepted that before humans started to settle in agricultural communities, they were nomadic. This had been the case for millions of years which means that sedentary communities are an extremely recent occurence on the time scale of human existence. For nomads, be they travellers on land or sea, their reference points are in the sky. Landscapes are constantly changing when you are on the move but skyscapes can be companions the world over. Anyone who walks for a few days in the mountains will witness the incredible presence of the night sky and the profound effect it can have on the psyche. There is no reason that this would have been any different 20 000 years ago, when homo sapiens was just as developed as now.

Nonetheless, when Dr. Chantal Jègues-Wolkiewiez, an ethno-astronomer, published her first paper on the astronomical significance of the Lascaux caves in Dordogne, France,[3] she received extremely negative criticism from archaeologists, and she still does. That humans were capable of reproducing the night sky during the paleolithic era is considered, strangely enough, to be far fetched. Putting forward strong evidence that the cave painting of a bull is actually a representation of the constellation of Taurus and that the group of six black dots above its neck is the Pleiades star cluster, could definitely be called "rocking the boat". For this would imply that there has been a continuity in the transmission of knowledge and symbol between the Lascaux cave painters and us.

The brightest star in the Taurus constellation is called Aldebaran which comes from arabic and means "the follower". It is the bull's eye, which is exactly where Chantal Jègues-Wolkiewiez places it on the cave painting (Figure 1). Two other stars in the

3   *Lascaux, vision du ciel des Magdaléniens.* Conférence ARCIVAM. Université Tous-Ages de l'Académie de Nice, 12/4/2001.

Figure 1. The Taurus constellation with the Pleiades above the bull's neck, in Lascaux caves, France (-18600).

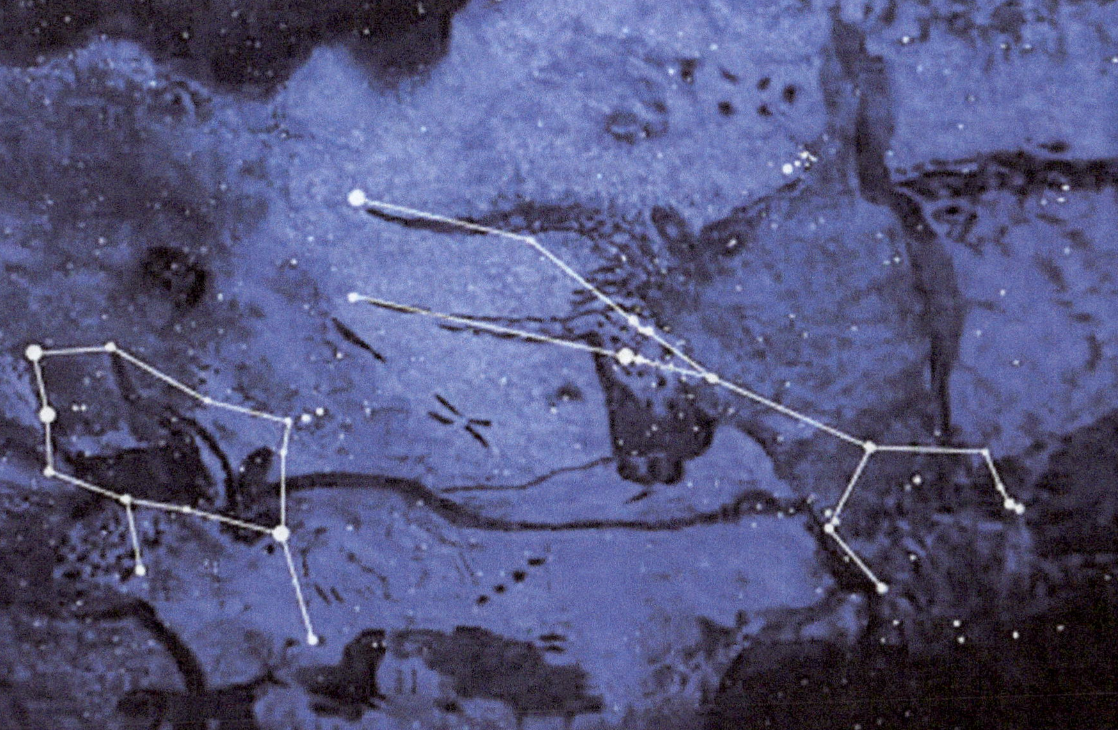

constellation are then in position at the tips of the bull's horns. She has also shown that the Lascaux cave was chosen because of its orientation. At sunset on summer solstice, the Sun would shine along the entrance corridor and right into the obscure chamber where the paintings are to be found.

This is where things start to get a little too complicated for people who have no direct knowledge of the sky and the movements of celestial objects. The word "solstice" may only be vaguely understood, corresponding to the longest or shortest day of the year, and may not bring any images into the mind's eye. It is not related to their own personal environment. Many people today have no real experience of the solstice and so do not realise how it could be important, seeing it as having superstitious origins in the pagan past, linked to Sun "worship". This vague idea that early religious behaviour was stupid with very little understanding and a lot of blind beliefs can be seen only as a projection of our own spiritual condition. Where else would the idea come from, seeing as we know absolutlely nothing about prehistoric thinking processes, spiritual experiences and belief systems?
What we do know is that in the most ancient civilisations, in Egypt and in Sumeria, students did not specialise but were asked to learn everything. Doctors in mediecine were also astronomers and priests, using geometry for architecture. Knowledge was seen to be unified, as epitomised by the universe, so it was to be approached from every angle, the central position being the seat of all wisdom. Music and harmony were not reserved for a specific type of person but seen to be based on number. This global approach to learning, which was a never ending process, is enormously different to modern specialisation, where academics know more and more about less and less.
The uninformed, and far too widespread, conception of the ancients being simple souls is a major obstacle facing anyone who wishes to present new ideas concerning prehistoric astronomy. It is not generally accepted that there is anything **to** be understood, apart, perhaps, the workings of a primitive mind. It is very difficult for western "civilised" adults to accept that there are gaps in their

own relationship to the cosmos and that this is a problem.

If humans discover in the future a way to travel along the wormholes in space and thereby transport a group of people to a far distant planet, a kind of futuristic Mayflower adventure, how would these people adapt to their new environment? What would they do to be able to feel at home? It seems quite obvious that, amongst other things, they would look to the heavens, count the number of suns and moons, see where they would rise and set and how long the different cycles would take. They would look for some way to find order in these movements, which would enable them to give structure to their own existence.

In modern western society on the planet Earth, many people are born, are educated, have children, grow old and die without ever having seen a moonrise. They may notice a massive full moon hovering over a building somewhere, but not once will they actually see it coming up from behind the horizon. Sadly, most people would have no idea where to look. They may never witness the extreme positions of the sunrises and sunsets, always believing that the Sun "rises in the East". Has this state of ignorance anything to do with the idea that the Sun is no longer "worshipped"?

The Borana tribe in Northern Kenya have a very interesting (and old) calendar system, based on the Moon. There are twenty seven different names for days (a western calendar uses seven). As this corresponds to the Moon's orbital cycle, seen from Earth it is the number of days between a moonrise at a same extreme position. For example, if we watch the moonrise day after day, we may see the point on the horizon where it emerges move towards the North. Then one day, the Moon will rise to the South of its previous position. This means that the day before, the Moon had reached its maximum northern rising point. From now on, it will rise further South every day until it attains its maximum southern rising point, from where it will head North again. It will come back to its most northern position after 27 days. This movement is regular and will repeat thirteen and one third times in a year or close to forty times in three years. So, in the Borana calendar, the name of the day indicates the direction of the moonrise on that day. It links spatial representation to calendar. It's as if Tuesday also meant East 15° North. When someone in the Borana tribe says the name of the day, they know where the moon will rise and set. They don't need books or astronomical tables or computers because they follow the Moon's movement every day. The name of the day also indicates the position of the Moon in the night sky constellations, which are named accordingly. This means that the zodiacal band, which we divide into twelve parts known as the signs of the zodiac, is divided into 27 parts by the Borana and each one coincides with the name of a day.[4]

The Borana month lasts alternatively 29 and 30 days, as in most lunar based calendars. This is because there are $29^1/_2$ days between two full moons.[5] They have twelve named months which make up the year. The first day of each month has a different name but it corresponds to the same **phase** of the Moon, the first visible lunar crescent after sunset. So the phase of the Moon is given by the number of days since the beginning of the month, the 13th day indicating the full Moon. Consequently, in this calendar, the name of the day, the number of the day in the month and the name of the month, our equivalent of Wednesday 3rd March for example, give all the information possible about the Moon; where it rises and sets, what phase it is in, what constellation it is crossing, but also the time of the year. By using this calendar on a daily basis, "ordinary" people link the Heavens and the Earth, remaining in contact with our closest neighbour, the Moon. They are not in any way Moon "worshippers", no more than Islamic or Jewish people who also use a lunar calendar. An arrogant interpretation of this kind can only come from ignorance.

The Borana calendar has another extremely important characteristic. The people who use it say it is linked to the stone alignments to be found on the edge of Lake Turkana and they have explained to ethnologists the alignments' relationships to

---

4   The ancient Chinese divided the zodiac into 28 parts known as mansions. As the lunar orbit lasts 27.32 days, it has been rounded up to a number dividable by 4. One of the mansions is, however, smaller than the others.

5   Many westerners do not know this and give 28 as the length of the lunar phase cycle, linking it probably to the menstrual period of 4 weeks.

certain stars. What is astounding is that these stellar alignments were exact in -300 but are no longer accurate today[6] which means that the information has been transmitted orally for over 2300 years without the slightest deformation. However, this change in star orientation does not affect the calendar because, despite what the informants of the Borana people told the ethnologists, the stone alignments are in fact pointing to summer and winter solstice and equinox sunrise positions which have only changed by a fraction of a degree since -300.[7] By observation along these alignments, the members of the community who are responsible for calibrating the calendar know when to add an extra month to maintain the relationship with the year cycle.

It has often been suggested that calendars were invented to enable early farmers to know when they should do certain things, like plough, sow or harvest. One of the major reasons for this idea is the appearance, in the doorways of cathedrals and churches of the early Gothic period, of zodiac signs linked to agricultural activities. For example, Pisces was associated with pruning, Aries with lambing and Libra with grape picking. These activities are, of course, linked to certain times of the year. So how do farmers who live in communities which use the lunar calendar know when to sow and prune? In fact, they do what farmers have always done, they observe what is happening in nature and act accordingly. The starting date for grape picking, for example, can vary by up to two weeks from one year to the next. People who work the land just **know** when they should do things. They don't really need people from cities to tell them what to do. The agricultural associations on the cathedrals were intended to work **the other way round**, showing people, who knew all about country life, at what times of the year the different zodiac signs corresponded.[8]

---

6   Because of the precession of the equinoxes.

7   The Sun's positions are not affected by the precession of the equinoxes but by a much longer and less accentuated cycle, the variation of the obliquity of the ecliptic.

8   As the zodiac signs are linked to the equinoxes and solstices, they are not affected by the precession of the equinoxes, unlike their counterpart constellations.

The different obstacles which prevent modern humans from understanding prehistoric astronomy (mindsets, lack of direct experience of celestial movement, sun worshipping and agricultural calendar theories, over-specialisation, misunderstanding of geometry and feeble symbolic thinking in general) make the task of writing about the Nebra sky disk a complicated one.

In this present work, the option has been chosen of putting the necessary "technical" or complementary information in appendices at the end of the book or, if possible, in footnotes, in order to simplify the main text and make for easier reading. However, to fully appreciate the beauty and power of the disc, an understanding of **geocentric** astronomy is necessary. It is also a path to reestablishing a long-lost relationship, that of Heaven and Earth.

Introduction

In 1999, near Nebra in the Saxony region of Germany, two treasure hunters made one of the most extraordinary discoveries in the history of archaeology. The object they discovered has come to be known as the Nebra sky disc and is the oldest recognised representation of the night sky, estimated to be some 3600 years old, dating it to 200 years before the first images in Egypt.

The implications behind the disc's existence were so enormous that the initial reaction of some authoritative specialists was to deny the disc's authenticity.

A German archaeologist, Professor Peter Schauer, of Regensburg University, claimed that the disc was a modern fake, and any idea that it was a Bronze Age map of the heavens was 'a piece of fantasy'. Professor Schauer stated that the supposedly Bronze Age green patina on the artefact had probably been artificially created in a workshop 'using acid, urine and a blowtorch' and was not ancient at all. The holes around the edge of the disc, he insisted, were too perfect to be ancient, and must have been made by a relatively modern machine.[9]

It was later revealed that Professor Schauer had never examined the artifact before making his claim, nor did he ever publish his findings in a peer-reviewed journal. Although the object has since been shown to be authentic, he refuses to accept the archaeo-chemical conclusions, suggesting that he is the spokesman of a silent faction of like-minded archaeologists. The majority of those who have studied it, however, recognise it to be a real eye-opener. *It was a thirty centimetre bronze disc covered with golden decorations. But the real sensation was that the golden decorations formed a picture, and this was something completely unheard of from the Bronze Age. It looked to me like the most significant archaeological find I'd ever seen.*[10]

9	Brian Haughton,*The Nebra Sky Disc, ancient map of the stars*, Ancient History Encyclopedia, www.ancient.eu.com ,2011

10	Dr. Harald Meller, BBC documentary, Secrets of the Star Disc, 29/01/2004

These differences of opinion between people who are supposed to give the "official" point of view show how sensitive this subject can be. It even brought one archaeologist to state:
We look like we are a bunch of crazies and don't know what we are talking about.[11]

In March 2006, an article about this disc appeared in the Times with the incredible title **"The Bronze age clock that told man it was spring."** This worrying phrase shows just how far removed from nature city dwellers have become, actually believing that prehistoric farmers would need a machine to indicate the seasons. Will the reality of the expression "Spring is in the air" be one day reduced to the level of the idiom?
In -1600, when the disc was buried, agriculture had been common practice in Europe for over 3000 years. The early neolithic site at Goseck (-4900), only 26km from where the Nebra disc was found, has been recognised as having an astronomical function,[12] indicating sunrise and sunset at winter solstice , the shortest day of the year. As it is thus accepted that astronomical observations had been taking place in the area for at least 3300 years when the Nebra disc was in use, is it not possible to surmise that these people were capable of a little more than knowing it was spring?
"We have been dramatically underestimating the prehistoric peoples."[13]

This *mea culpa* is heartwarming, despite the use of the Royal "We", but there is evidence that this "dramatic underestimation" has been dramatically underestimated and that these prehistoric people were much more advanced than we dare imagine, perhaps more so than us.
It's the find of a lifetime, indeed the find of several lifetimes. What it's doing is making people think for the first time, a society that

11   Christian Wunderlich, Saxon Anhalt state archeochemist in Archaeology Magazine, December 2005.

12   Ulrich Boser - Solar Circle (Archaeology Magazine July/August 2006)

13   Dr. Harald Meller quoted in the article, *Bronze Age clock that told man it was spring* by Roger Boyes, The Times, March 2 2006

can make this is complex, it's sophisticated, it's intellectual.[14]

In fact, a lot of people had thought this for a long time before the discovery of the Nebra sky disc. In 1897, Felix Gaillard from Plouharnel, near Carnac in Brittany published "Prehistoric Astronomy"[15], which included a statistical study of 220 megalithic dolmens in southern Brittany. This showed that not one single dolmen had its opening between due West and North-east, an angle of 135°. Gaillard concluded that their orientation had not been left to chance, nor was it linked to topographical considerations but that it was more probably determined by astronomical factors and in particular the Sun's rising and setting positions. This book has been copiously ignored by mainstream archaeology ever since and does not even figure on the list of Gaillard's works in the French Encyclopedia of Authors.

Sir Norman Lockyer, founder of the magazine Nature and inventor of the science of astro-physics, being the first person to detect the presence of Helium in the Sun, was later ostracised by the "scientiflc community" for his belief in advanced astronomy in neolithic and Egyptian civilisations. He thought that the pyramids and temples of ancient Egypt were constructed

"in strict relation to the stars, and that by applying a knowledge of astronomy to the actual site orientation of the various temples, accurate dating can be achieved."

Another non-archaeologist academic who naively thought his work could be of interest to stone age specialists was Alexander Thom, Professor of Engineering at Oxford University. He went the whole hog and not only suggested that neolithic man was an accomplished astronomer capable of predicting eclipses but also that he had used Euclidian geometry, Pythagorean triangles and had a precise unit of measurement, used all over Great Britain and Brittany. Although his methodology and results were accredited by the Royal Society of Statisticians, and despite a brilliant BBC

---

14   Professor Miranda Aldhouse Green, BBC documentary, Secrets of the Star Disc, 29/01/2004.

15   Félix Gaillard, Astronomie Préhistorique, 1897, reedition Epistemea, 2004.

Chronicle documentary which clearly presented his "stone age Einsteins", his conclusions have been far too difficult to accept for the mainstream archaeologists who have managed to delete him from all university courses, including the Archaeo-astronomy course at Leicester University.[16]

The list of people who have produced important work on this subject is far too long to be reproduced here. Most of them have been upset and frustrated by the categorical rejection of their contributions and must simply remain flabbergasted when they hear that *"people are thinking for the first time."*

In 2006, Ralph Hansen, an astronomer from Hamburg, presented a new theory concerning the Nebra sky disc. He suggested that it was

an attempt (sic) *to coordinate the solar and lunar calendars. It was almost certainly a highly accurate timekeeper that told Bronze Age Man when to plant seeds and when to make trades, giving him an almost modern sense of time.* [17]

Mr. Hansen comes to these conclusions from the size of the crescent moon, which is not a new Moon but is four or five days old. He consulted the "Mul.Apin" collection of Babylonian documents, dating from around -700, and in particular, it can be deduced, the second tablet on which there are two ways of determining when to insert an additional month in order to keep the lunar and solar calendars in phase. One method uses the rising dates of certain stars while the other uses the position of the moon in relation to the stars and constellations.

It is interesting to note that this second tablet also gives many other astronomical indications including the relative duration of day and night at the solstices and equinoxes, and the lengths of shadow cast by a gnomon at various times of the day at the solstices and equinoxes. These latter techniques were most probably known to the builders of the Goseck circle.

16   For more details, see «*Alexander Thom, Cracking the Stone age Code*», Robin Heath, Bluestone Press, 2009

17   *Bronze Age clock that told man it was spring* by Roger Boyes, The Times, March 2 2006

It is believed the Bronze Age astronomers compared the Nebra clock with the sky. The 13th month was inserted when the sky corresponded with the map on the disc. This happened every two to three years.[18]

Who believes this? Most probably people who have watched too many television documentaries where fact and fiction are intermingled to such an extent that it is impossible to know which is which. How did the person who made the disc know what he was doing? The knowledge must have existed **before** the disc and it was most certainly the people who had that knowledge who decided when an extra month should be inserted into the calendar, not some mindless idiot who was vaguely trying to compare a clearly symbolic representation with the reality of the night sky. And what would he do if it was cloudy? Go out to the local farmers and say "Sorry chaps, filthy weather conditions. You'll have to wait for a month to know if it's time to sow."

The sensation lies in the fact that the Bronze Age people managed to harmonize the solar and lunar years. We never thought they would have managed that.[19]

In this present work, it will be shown that the object is not merely a symbolic representation of the night sky but a device which allows its owner to keep track of and accurately predict, not only Sun and Moon cycles, but also conjunctions of Mercury, Venus and Mars at the very least. It will become clear that the people who devised this object had a much greater knowledge of the Heavens and its cycles than is currently accepted for this period. They practised precise complex geometry. They had even deciphered an underlying coherence in the solar system, based on the number 39. Although this information was punched out in holes around the Nebra sky disc, *"too perfect to be ancient"*, it has been totally ignored by all the specialists who have examined the object.

----

18   *Sky disc revelation shines light on Bronze Age*, Allan Hall, theage.com. au, Berlin, March 1, 2006

19   *Ibid*

But the German researchers also discovered that in the 400 years that the disc was in use, its status changed. The perforations on the edge, as well as the addition of a ship to the map, suggest the knowledge about the lunar calendar was lost.

"In the end, the disc became a cult object," Dr Meller said.[20]

How unfortunate, and incomprehensible, that degenerate descendants should have mutilated such a precious object! Unless of course, these later additions were improvements to the initial project, which is what is normally supposed to happen as science progresses.

Other researchers justifiably doubt its use as a sky **observation** artefact, putting into question the precision of its making by showing that the two peripheral Sun arcs are not on the same axis. *"I do not think it was used as an instrument used for observing objects in the sky. I can't find any evidence for this." Roslund and Pasztor argue that few features on the disc tend towards exact representation and that it is more likely to have been of symbolic value - perhaps used in shamanic rituals.*[21]

The symbolic (and geometrical) nature of the disc is self evident. The introduction of (non-intellectual) Shamans is, regrettably, an all too common way to explain away the unexplainable, without, of course, explaining anything at all. Moreover, the confusion between symbolism, a mode of expression using form, and shamanism, magico-religious practices using altered states of consciousness, shows an embarrasingly scanty understanding of ancient knowledge and its communication. Unfortunately, this analysis has been repeated by other researchers, and in particular Dr. Euan MacKie, one of the few remaining defenders of Professor Thom's work, who elsewhere has argued in favour of the existence of a prehistorical astronomer-priesthood.[22]

---

20   *Ibid.*

21   Curt Roslund of Gothenburg University, BBC News,25 June 2007

22   Dr. Euan MacKie, *The Prehistoric Solar calendar: An out-of-fashion Idea Revisited with New Evidence,* Time and Mind, March 2009, p.28-30

The most spectacular thing to see is the final piece in the jigsaw. And these are the Pleiades. [23]

As the final piece of the jigsaw has already been placed, it is not certain that there is any room for this present work. Where on Earth could one put it? The reader will have to make up their own mind about the answer to this question.

----

23   Prof Miranda Aldhouse Green (University of Wales), ibid.

BASIC FACTS

The Nebra sky disc[24] is a bronze disc of around 32cm diameter. The discovery site is a prehistoric enclosure encircling the top of a 252 metres (827ft) elevation in the Ziegelroda Forest, known as Mittelberg ("central hill"), some 60km west of Leipzig. The surrounding area is known to have been settled since the Neolithic, and Ziegelroda Forest is said to contain around 1,000 barrows. The pit in which the disc was found contained other objects, including two swords, two axes, chisels and armlets. The disc has been associatively dated to -1600. It underwent an exhaustive battery of tests that appear to support the artefact's authenticity.

The copper in the disc has been shown to have been extracted from bronze age mines in Bischofshofen in Austria, while a recent analysis found that the gold was from the river Carnon in Cornwall. The tin content of the bronze was also from Cornwall. The disc was given a deep blue patina by smearing it with rotten eggs, causing a chemical reaction on its bronze surface. This would have simulated the night sky. Ageing has changed the colour to its present blue-green. A mixture of hard crystal malachite now covers the artefact. The large size of the crystals can only be caused by a slow process, eliminating the possibilty of some accelerated method using urine and acid, as had been suggested. Once the night sky colour had been obtained, gold leaf was then used to add symbols which can be seen in Figure 2. Thirty two gold dots are thought to represent stars. Seven of these are arranged in a group, just above and between the two major central symbols, a circle thought to represent the Sun and a crescent shape resembling a Moon phase. The seven star cluster could represent the Pleiades.

Two golden arcs along the sides, which have been shown to be later additions, are symmetrical. They each open an angle of 82° around the perimeter. One of the arcs has been placed over two gold dots, while it would appear that, before the other arc was positioned, a gold dot was moved. Another addition was a smaller

---

24   These facts have been collated from all the reliable sources available to the author.

Figure 2. The Nebra sky disk

arc, at the bottom of the disc if the "Pleiades" are placed towards the top, surrounded with multiple strokes and which has come to be known as the Sun Ship. Finally, thirty nine holes were punched in a regular fashion around the disc's circumference.

It has been suggested that the golden arcs on the perimeter represent the yearly movement of the sunrises and sunsets along the horizon, as seen from the Mittelberg, where the Nebra disc was found. The angular value of 82° accurately corresponds to the winter and summer solstice positions at this latitude.[25]

---

25   For an explanation of the Sun's angular movement along the horizon during one year, see "Appendix 1 : Solstices", page 60

Solar Geometry

The Mittelberg (252m) is a panoramic viewpoint. The Brocken mountain, the highest in Germany at 1141m, is at an angle W42.82°N from there, which is too far North by 2.5° for a midsummer sunset in 1600BC (W40.38°N), as was suggested by certain researchers. However tempting it may be, the angular difference is too great to accept a correlation, as the Sun's size on the horizon is about 0.6°, meaning it would have set about 4 sun diameters to the south of the Brocken mountain peak.

The Wurmberg (971m) is the second highest mountain in the Harz and the highest in Lower Saxony. It is about 5 km due South of the Brocken at an angle of W40.5°N, so it is a much better candidate for the midsummer sunset as seen from the Nebra disc site. Between the two, the Kalte Bode valley runs in the direction of Nebra. At an earlier period, around -5000, which is the date of the nearby Goseck site, the sun would have set, at its most northerly point (W41.2°N) in a 41000 year cycle, between the two peaks when seen from Mittelberg. Its rays would then have shone down the Kalte Bode valley.

The fact that the tin and some of the gold used to make the Nebra sky disc was mined in Cornwall would imply that the local people were in contact with the dwellers of southern Britain. The astronomical elements inscribed on the disc could easily be part of the same cultural knowledge that lay behind the building of Stonehenge, especially considering the fact that the two locations are on a very similar latitude and so astronomical observations would be equivalent.

The precise internal geometry of the disc is worth closer inspection. Firstly, in Figure 3 the disc has been orientated so that the Pleiades, the cluster of seven gold dots, is in the "sky" above and between the Sun and the Moon. The top of the disc is to be considered as the South, as the Pleiades are visible in the Southern sky. The lower part of the disc then becomes the Northern sky. This is the typical orientation of the sky in ancient Chinese representations and also the Pa Kua, the basis of Feng Shui (Figure 4).

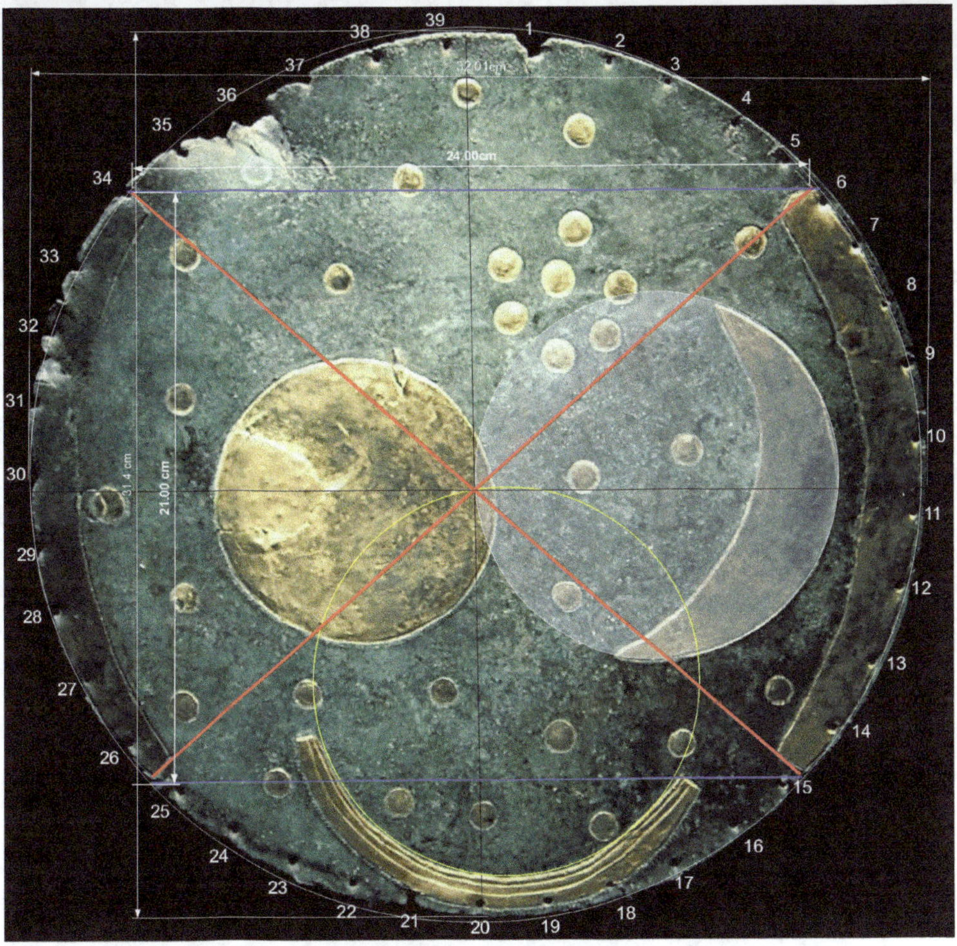

Figure 3. The Nebra solstice rectangle and its central point.

Figure 4. The Pa Kua

Figure 5. The Crucuno Quadrilateral

As a result of this orientation, the Sun is on the left of the disc and the Moon is on the right. In a previous work, the present author has shown this same organisation in dolmens around Carnac[26] and even on the main porch of the Gothic cathedral in Laon, Northern France. This symbollic representation is chosen because, when

____

26   Howard Crowhurst, *Mégalithes, Principes de la première architecture Monumentale*, Edition Epistemea, 2007.

26

one looks to the East, the Moon, at its **major** Southern rise[27], appears to the right of the extreme Sun position at winter solstice.

The disc has been positioned precisely so that a line (in blue) joining the upper ends of the golden solstice arcs is horizontal. We discover that the blue line joining the lower ends is also exactly horizontal. The lines in red, which join the opposite extremities of the golden arcs, cross at a point just inside the right border of the sun disc. This exact point is also touched by the perimeter of the Moon circle, in white, the inner circle of the "sun boat" (in yellow) and the vertical diameter of the whole disc (in black) which also bisects two golden dots and hole 20, indicating true North. The exact centre of the disc is slightly higher than this junction point. This structure shows the precise (and intentional) organisation of the different elements and explains the general harmony of the ensemble.

As mentioned, the golden solstice arcs mark an angle of 82°. How was this measured?

A megalithic monument near Carnac in Brittany may give us a clue to this enigma. The Crucuno quadrilateral (see diagram and aerial photograph Figure 5) is a rectangle made of standing stones and orientated to the cardinal directions. The shorter side of the rectangle is North-South. At solstices, the sunrises and sunsets take place along the diagonals of the rectangle, shown in red in the diagram.

The astonishing fact about this monument is that the proportions of the sides of the triangle are 3 to 4. In other words, the solstice angle from east at Carnac when the monument was built some 6000 years ago corresponded to the angle of a 3-4-5 triangle, 36.87°. The total solstice angle between extreme positions is twice this, 73.74°. The same principle can be seen to have been used in many monuments in and around Carnac. This cannot be a coincidence and must suggest that the builders had made a link between astronomy and geometry.

---

27   See page 44 and "Appendix 3 : The Moon's 18.6 year cycle", page 62

At the latitude of Nebra, further North, the overall solstice angle is greater by a little over 8° if we accept the angle of 82°, which is that shown on the Nebra sky disc. This corresponds to the diagonal of a rectangle one unit larger than the one seen at Crucuno; 7 by 8 units at Nebra rather than 6 by 8 at Carnac.

The angle of the diagonal of a 7 by 8 rectangle is 41.19°, so the overall solstice angle where the diagonals cross is twice this, 82.38°, which is very close to the maximum solstice angle at this latitude. The measurements on the diagram show the rectangle to be 21cm high by 24cm wide, an exact 7 to 8 relationship.

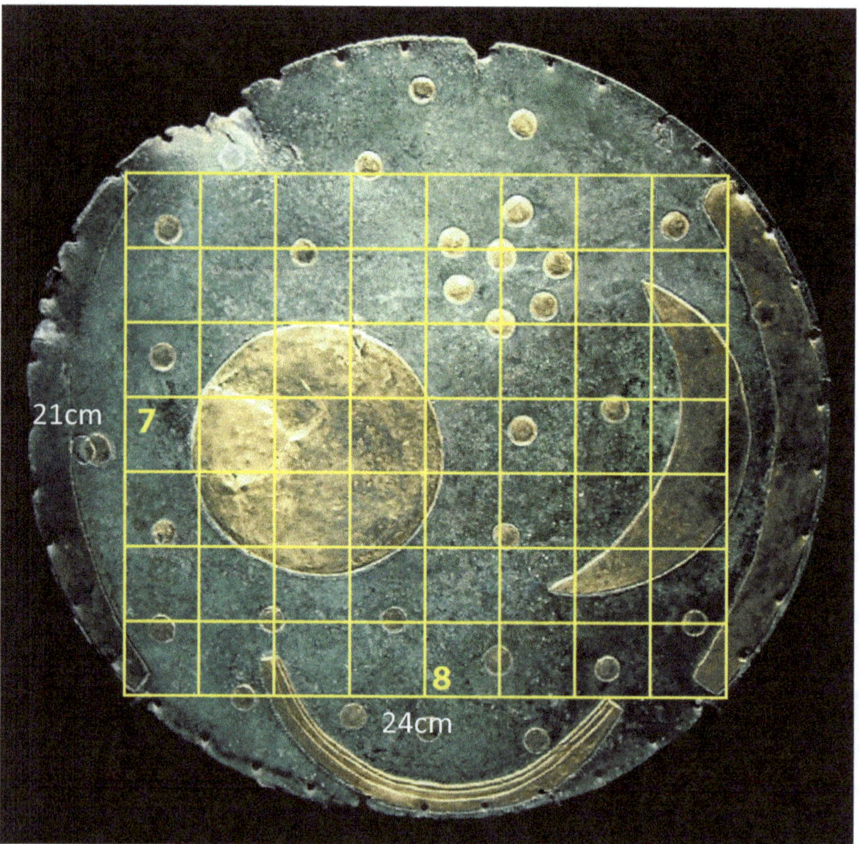

Figure 6. The 7 to 8 solstice rectangle

THE PLEIADES AND MAY DAY

The third mountain top on the horizon to the north west of Mittelberg is the Kyffhäuser. At an angle of W25.2°N, it is perfectly positioned for a sunset on the May 1st in -1600.

However, a tradition still known to this day links the Brocken to May Day. In Germany, Walpurgisnacht, the night from April 30th to May 1st, is the night when witches are reputed to hold a large celebration on the Brocken and await the arrival of spring.[28]

Could there be some other site, placed along the May Day sunset line from the Brocken, waiting to be discovered?

The representation of the Pleiades star cluster on the Nebra sky disc is an incredible confirmation of the date of its fabrication since May 1st is also the date of the heliacal rising of the Pleiades in **-1600**.[29]

The heliacal rising of a star is an important moment because it allows an exact determination of the year length. It is the moment when the constellation appears on the horizon just before sunrise. In Egypt, the heliacal rising of Sirius, the brightest star in the sky, fixed the beginning of the year. Solstices, where the sun rises at the same place for several days, are ideal for marking out the land and positioning the extreme positions of the movement of the zodiacal band. They are more approximate for calendar purposes and require organised observatories to make precise measures. The heliacal rising, on the other hand, can be observed from any spot with a clear view to the East. It is not necessarily a position one is seeking but a particular star, so it's easy if you know what you are looking for. When you can see the star for the first time just before sunrise, then you know it's a particular day of the year. The Pleiades cluster, which is easily recognisable, is thus an ideal choice.

Historically, May 1st seems to have a relationship with the Pleiades. The seven stars were often associated with virgins, young women and purity. The brightest star in the constellation was called Maia

---

28   Oxford Dictionary of Phrase and Fable, Edited by Elisabeth Knowles, Oxford University Press, 2005

29   See "Appendix 7 : Calculating the Heliacal Rising and Setting of Pleiades", page 70

by the Romans, the Latin name of the month of May. She was the mother of Hermes (Mercury), the winged messenger and god of communication. Hermes was also the god of high places, for he was born in a cavern on a mountain top, and as such he is often equated with St Michael, the dragon tamer. In Chinese, the name for the Pleiades is Mao and in South America, May (the Mayan civilization was a calendar based religion). In many folk festivals in Europe the May Queen is chosen. She is a beautiful girl who is not married. Processions take place where men disguise themselves as horses (Obby Oss in Padstow, Cornwall) and dance in the streets, like the dancing dragons in China. Up until recently in France, on May 1st young men visited each house where an unmarried girl lived, and hung a bouquet of flowers on the gate. May 1st has been important for people all over the world for thousands of years. The Nebra sky disc provides evidence that this link between May Day and the Pleiades is very ancient and probably was initiated in -1600 when the heliacal rising of the Pleiades fell on May 1st.[30]

30   See "", page 66

Figure 7. Queen Guinevre's maying by John Collier (1900)

THE THIRTY NINE STEPS.

One feature of the Nebra sky disc which seems to get little attention is the presence of holes around its perimeter. According to Wikipedia (in 2012):
By the time the disc was buried it also had thirty-nine or forty holes punched out around its perimeter, each approximately 3 mm in diameter.

39 or 40 is a very approximate estimation for an object that has been thoroughly examined. This is probably because the researchers would like the disc to have 40 holes, a nice round number linked to 4 and 10, but in reality, it only has 39 holes and 39 is not a number recognized as significant. The uncertainty comes from a small part of the perimeter which is missing (between holes 35 and 37) and the slight distortion of the edge between holes 31 and 32 and holes 32 and 33.

One only has to look at the spacing of the holes all around the disc to see that, even if it has not been executed precisely, the holes have been positioned with regularity. A close examination shows that none of the holes are directly opposite each other on the perimeter; the point opposite is approximately half way between two holes, which is always the case when a circle is divided by an odd number of points. This must have been done intentionally, as it is much easier to place points opposite each other. This would suggest that the **number** of holes is important but their precise spacing is less so, again adding weight to the idea that the disc was not an observational device. It would, however, be impossible to cram two holes in between holes 35 and 37 without creating an anomaly, and so it must be recognised that the correct number of holes is 39.

To the author's knowledge, no interpretation has been given of the possible meaning of 39 holes on a sky disc. It is suggested that they are an irrelevant afterthought, showing that the initial understanding of the sky disc had been lost. What is the evidence leading to this hypothesis? This is not the only case in prehistory of the division of a circle by a high number. The archaeological excavations at Stonehenge have shown the high precision spacing of the 56 Aubrey holes on a perfect circle just inside the outer trench, "punched out", in this case, much earlier than the better known central monument.

The first thing one must realise is that **it is not possible to space 39 holes around the circumference of a circle by accident.** Then, of course, one wonders **how** it was done and **why.**

Looking first at how it was traced out brings to light some very interesting results.

In a circle, there are 360 degrees. If this is divided by 39, we obtain 9.23°. Nothing special here, apparently. However, when we multiply this angle by 2, the result is 18.46° which is only 0.025° away from the acute angle formed by the diagonal of a triple square. This angle, when doubled again, gives the smallest angle of a 3-4-5 triangle 36.87°. The geometrical method using a 3-4-5 triangle and a triple square to trace the holes is shown in "Appendix 2 : Dividing a circle into 39 equal parts", page <?>.

In a previous work, the author has shown how these two angles measured from East gave the direction of the two major megalithic alignments in Carnac, known as Le Menec and Kermario[31]. We will see later how they were used to lay out the megalithic complex in the Boyne Valley in Ireland, which includes Newgrange.

The question of **why** 39 holes were punched around the circumference remains. Much discussion has taken place between researchers as to why there are 32 gold dots on the disc and various interpretations have been put forward, all of which show the relationship of 32 to heavenly cycles. Now the gold dots are not arranged in any regular way. Six of them form the Pleiades constellation, seven others are in the zodiacal belt, in the width of the Sun and the Moon, a further seven are above the belt, six more are in a slightly winding line below the belt, three are in the golden arc known as the Sun Boat and a final dot is in line with those three but just outside the "boat". This organisation would in no way facilitate counting a cycle of 32 units. One of the advantages of holes, compared to gold dots, is that one can put pegs in them to mark positions and if one wants to count cycles,

---

31   Howard Crowhust, *Carnac, The Alignments*, Epistemea, 2010

putting regularly spaced holes around the perimeter would seem to be a much more practical solution than irregularly placed gold dots. As we are sure it is a sky disc, the obvious question is "Are there any cycles linked to the number 39?" and the answer is "Yes".

MARS AND THE MAYAN

The most obvious cycle based on the number 39 is the synodic cycle of Mars which is 779,9643 days long. This means that every 780 days, which is 20 times 39, Mars comes back to exactly the same position in relation to the Sun. In other words, between two alignments Sun, Earth, Mars, there are 780 days. So if a red (Mars) peg is placed in a hole in the disc the day that Mars culminates (at its highest point in the sky due South) at midnight (when the Sun is due North) and that peg is moved by one hole every 20 days, then after 1 rotation, Mars will be back in the same place at midnight. After half a "lap", the Sun and Mars will be in conjunction. In fact, the peg's position on the disc will give its exact position in relationship to the Sun. If Mars is positioned as suggested, it would indicate the central point of its retrograde movement (see Figure 8).

It could however be positioned at any given moment, such as its conjunction with Venus, or the Moon.
Now, is there a system for measuring 20 days on the disc ? Again the answer is "yes". This can be done using the golden arcs along the sides of the disc which have been seen to represent the Sun's

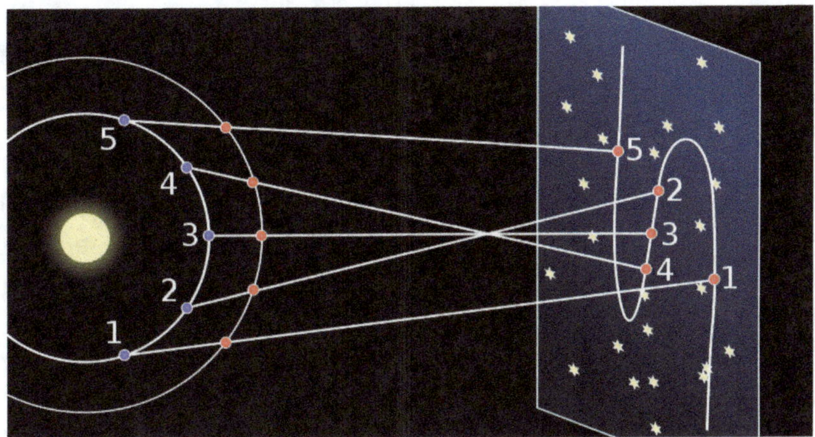

Figure 8. The retrograde movement of Mars (outer circle) during the period of its solar opposition seen from the Earth (inner circle).

movement through the year. The holes divide these arcs into 9 segments. If each segment has an angular measurement of 9.2°, 9 segments totalize 82.8° which has been seen to be the Sun's angular movement on the horizon between its extreme positions at solstices. As it takes the Sun half a year (180 days) to go from one extreme to the other, each segment can be seen to mark 20 days (180 divided by 9). This is not an exact representation, as the Sun's movement along the horizon is not regular but it would be symbolic and precise at solstices and equinoxes. After the 2.13 year Mars cycle, because the year lasts 365 days and not 360, the Sun peg would be incorrect by about 10 days or half a segment. After two Mars cycles, this would become one whole segment and so the Sun peg would be moved back a hole to "put the clock right".

It is perfectly established that the Mayan civilisation, well known for its astronomical and calendar sciences, had a vigesimal number system (based on 20) and one of its basic time cycles was the *uinal*, which was 20 days long. Each day of the cycle had a name, making 20 day names. There was also a 13 day cycle, called the Heaven God cycle or nowadays the *trecena* cycle, where each day was identified by a number. 13 *uinals* made up the 260-day *tzolkin*, where each day was a unique combination of a name and a number. So, for example, 1 *Imix* was followed by 2 *Ik*, 3 *Akbal* and so on. This was the basis of the sacred calendar that determined the ceremonies and prophesies of the Maya.[32] 260 days is one third of the 780-day Mars cycle.

13 is one third of 39 so if a Maya peg was moved by one hole per day, after one lap of the disk it would have completed three of the Mayan *trecena* cycles. After 20 laps, it would accomplish 60 *trecena* cycles, 39 *uinals*, 3 *tzolkins* and 1 Mars cycle. We shall see shortly that this astounding connection between the Nebra sky disc and the ancient Mayan calendar can be taken further.

The remains of a vigesimal counting system in Europe can be found in the word "score", *scoru* in Old English meaning twenty, but also to mark or tally, *skor*, in Old Norwegian, which is the origin of the

---

32   Geoff Stray, *The Mayan and other ancient Calendars,* Wooden Books, 2007

modern sense in sports, the scorer in cricket being the person who marks the number of runs. Its use for measuring time is found in the expression "three score and ten", meaning the seventy years of life expectancy. In modern French, eighty is still called *quatre-vingt*, four-twenties, a remainder of the old system.

A forty day period, twice twenty, was mentioned by Hesiod[33] for the Pleiades heliacal cycle. He linked it to key moments in the agricultural calendar. The heliacal rising indicated the moment to harvest and the heliacal setting was the time to plough the fields. This led Professor Miranda Aldhouse Green to state the following: *We know from Greek writers that the Pleiades were used as an agricultural marker, so that farmers knew when they should do certain agricultural activities. So what the Nebra disc does is to tell people not only the right time to do it but it is the blessed time to do it.*[34]

This a priori conclusion would appear impossible. If the Pleiades cluster had been used for agricultural reasons in Greece in -700, it could not have had the same function in Germany in -1600. It is not at all the same climate and the German farmers would have had to harvest **earlier** than their Greek counterparts (May 1st instead of May 13th).[35] Hesiod may imply a symbolic, rather than literal, meaning for his text. The forty day period was considered to be a cycle of transformation, as attested in the Bible. The present-day period of Lent still maintains this principle.

Forty days also separate the equinox from the 1st May. This period corresponds to the Sun's regular movement on the horizon, on either side of the equinoxes, before it starts slowing down as it approaches the solstices. It is possibly shown on the left hand side of the disc. One of the golden dots, which is claimed to have been moved by about half its size towards the centre, marks the central position of the golden arc, the position of the equinoxes. As there are two equinoxes, Spring and Autumn, the double circle may be seen to be intentional. This would be confirmed by the fact that

---

33   Hesiod, Works and Days (ll. 383) When the Pleiades, daughters of Atlas, are rising, begin your harvest, and your ploughing when they are going to set. Forty nights and days they are hidden and appear again as the year moves round, when first you sharpen your sickle.

34   Professor Miranda Aldhouse Green, *ibid.*

35   "Appendix 7 : Calculating the Heliacal Rising and Setting of Pleiades"

two dots on the right hand side of the disc were not moved when the golden arc was added, but covered over.[36] These two circles are exactly aligned along the diameter of the central golden Sun disc (marked by a dotted yellow line on Figure 9). A golden dot was placed two segments above and two segments below the double circle, thus marking a 40 day period on either side of the equinox (marked by green lines). The higher dot would indicate sunset on the 1st May (behind the Kyffhäuser mountain) and the lower one could indicate the beginning of the 40-day period leading up to Easter, now known as Lent.

If the golden arcs were added later, as suggested by the analysis of the gold, it may have been to reinforce the solar meaning of these dots which had been present from the outset.

36   As one is golden and the other is not, they could also represent the Sun and the full Moon and thereby indicate the feast of Easter which is the full moon following the Spring equinox.

Figure 9. The forty day periods around equinoxes.

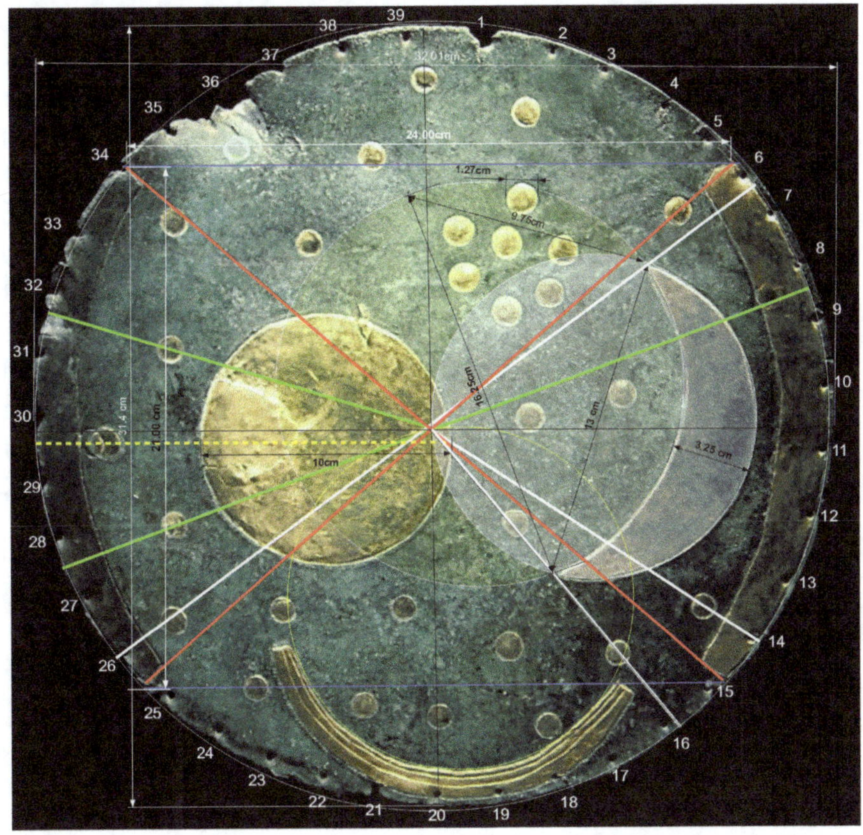

Venus, the Moon and Mars

The next cycle closely linked to the number 39 is Venus, whose synodic cycle is 583.92 (584) days. This number is one removed from 585 which is 15 times 39. Venus, being closer to the Sun than the Earth, is never in opposition in the same way as Mars. When it is at its furthest from the Sun, its angular distance is 47.8°, which is a little more than one eighth of the 360° zodiacal belt. When it is observed from the Earth as the Morning Star, Venus will rise earlier each day, seeming to move away from the Sun until it reaches this extreme position where it will stop for a short time, just as the Sun does on the horizon at solstices. Then, it will slowly move back towards the Sun, rising later and gradually become more and more difficult to see until it becomes invisible, its rising being obscured by the light from the rising Sun. It will remain invisible until one evening, when it will just be visible near the horizon shortly after sunset. Again, it will set later and later, moving away from the Sun once more, but this time on the other side of it. As the Sun sets, it will be higher in the sky as the days go by until, again, it comes to its extreme elongation of 47.8°. If it is in conjunction with the Moon at that time, the Moon would be lit in the same way as it is on the Nebra sky disc, the central part of its crescent measuring one quarter of its total diameter.[37]

If the Venus peg was positioned at the time of a conjunction with the Moon in this phase after sunset, an easily visible and recognisable event, and moved one hole every 15 days, then it would come back to its extreme position in relation to the Sun one rotation later. After half a lap, it would be at its extreme position on the other side of Sun, before sunrise, when it is known as the Morning Star. After one quarter and three quarters of a lap, it would be in conjunction with the Sun and consequently invisible.

---

37   It is interesting to note that the way to draw a crescent moon of this shape is by using a circle which has a diameter one quarter greater than the diameter of the circle of the Moon. This can be done using a 3-4-5 triangle (see Figure 9). Since the Sun and the Moon have the same apparent size in the sky, the shape of the crescent on the sky disc excludes the possibility that it may represent an eclipse.

The existence of a 15 day period can be confirmed by the fact that it has a name in certain languages; *cóicthiges* in Old Irish, *pythefnos* in Welsh, *quinzaine* in French. The ides of March (the date of Caesar's assassination) was the fifteenth day of the month in the Roman calendar. It is half of the solar month of 30 days, 12 of which make up the 360 day calendar, as used in Ancient Egypt. The name of Venus is also of interest. Its origin is said to be the PIE root *wen-* meaning to desire which led to the Latin goddess of beauty and love. The exact name Venus exists in Icelandic, Norwegian, and Welsh. However, *wen* in Welsh means white and since Venus is the brightest star in the sky, visible sometimes in broad daylight, the connection seems possible. One of the Greek names for Venus was Phosphorus which means light and which later gave the name to the white element, used in matches, which bursts into flames. If Venus means white, then the link with the *Veneti*, the druidic people living in Carnac before Caesar's invasion, becomes a possibility.

For every four circuits of the Venus peg around the Nebra sky disc, the Mars peg would do three since 4 times 585 equals 3 times 780 equals 2340 days. As the Venus cycle is 584 days long in reality, 4 Venus cycles would make 2336 days, 4 days shorter than 3 Mars cycles. This means that a certain Mars Venus conjunction would be repeated between 2336 and 2340 days or 6.4 years later (see "Appendix 4 : 1st May -1602", page 63). 5 of these Mars-Venus conjunctions (11690 days) are 2 days longer than 32 years, which in turn is equal to 4 times the 8-year Sun-Venus-Earth inferior conjunction cycle (2921 days).

For example, if the Nebra sky disc had been set at the Mars-Venus conjunction, 3° from the Sun, on April 3rd 1949, the precise Mars-Venus-Sun-Earth alignment which took place on April 5th 1981, 32 years and 2 days later, could have been predicted. The next conjunction in this cycle happens on April 7th 2013.

Could this explain the presence of 32 dots on the disc?

THE SUN, THE MOON AND GESTATION

The most astonishing connection between the Nebra sky disc and natural cycles is linked to the period of 273 days which is 39 weeks of 7 days. This corresponds to 10 solar rotations, 10 lunar rotations and also 10 lunar orbits. The Moon is said to be in synchronous rotation, meaning that it turns on its axis at exactly the same angular speed as its movement around the Earth. This is why we always see the same face and why the Moon has its "dark side" which is never visible from Earth. The fact that it turns at the same speed as the Sun, when seen from Earth, is not particularly well known. 273 is consequently the most important whole number which links the Sun and the Moon, the two dominant symbols on the sky disc. Furthermore, 273 days is considered to be the average length of the human gestation period. It is said to be 9 months in Western countries. These are nine solar months of 30 days. In China, pregnancy is said to last 10 months which are 10 solar and lunar rotations of 27.3 days.

273 is 260 plus 13 and this brings us back to the sacred Mayan calendar. Following on from what we saw on page 36, 273 days is 1 *tzolkin* plus 1 *trecena*. Now the 20 day names of the Maya can be presented in two different orders. The first and most obvious way is their daily succession in the calendar. The second way is their succession as the first day of a *trecena*. For example, if we start with 1 *Imix*, then one *trecena* or 13 days later is 1 *Ix*, then 13 days after that 1 *Manik* etc. So 273 days after 1 *Imix* is 1 *Ix* and 273 days after that is 1 *Manik* etc. The cycle would come back to its initial position after 5460 days, which is 20 periods of 273 days but also 21 *tzolkins*.

It is the present author's belief that this is the first time that the existence of this solar-lunar gestation cycle in the Mayan calendar has been identified.

The importance of the 273 day cycle was well known in medieval times. In the Vézelay Basilica in France, dating from the 12th century, a zodiac sculpted in a semi circle around Christ in Glory, (Figure 10), is composed of 27 circular medallions plus one third of a medallion making a total of 27.3. In this crescent **moon shaped**

Figure 10. The Vezelay Zodiac in France. The one-third size medallion has been enlarged on the left.

form, the image of a stork is represented, clearly showing the link with childbirth. Another example is the famous labyrinth in the floor of Chartres Cathedral where 273 paving stones form the path which leads into (or out from) the centre of the maze. Finally, Michaelmas, or Saint Michael's day, one of the quarter days in England, is celebrated on 29th September in the Roman Catholic Church.

 This is the 273rd day of the year (in leap years). The quarter days divided the year into 4 periods of 91 days. 273 days is equal to 3 of these periods, showing its 3 to 4 relationship to the year.

By moving a Moon peg around the perimeter of the Nebra sky disc by one hole per week, after one lap, the moon would be in precisely the same position in relation to the stars (and the zodiac) as it was initially. The Sun would also be showing exactly the same face. For every 4 moon peg moves, a Sun peg could be moved by 3 holes. It would then do one lap in a year less a day (52 weeks or 364 days).

LUNAR GEOMETRY

The monuments in the Boyne valley in Ireland, especially Newgrange and Knowth, can give more information concerning the use of the thirty ninth division of the circle and the associated geometry using the triple square and the 3-4-5 triangle. These two monuments are linked by a special (minor southern) moonrise angle.[38]

"The angle between the centres of two of these enormous neolithic mounds, from Knowth to Newgrange is E 34.69°S (Figure 11). It is known to correspond at this latitude to the minor southern moonrise which occurs every 18.6 years. It corresponds exactly to the angle of the diagonal of a septuple square added to the diagonal of a double square.

$$8.130° + 26.565° = 34.695°$$

which is also the diagonal of a 9 by 13 rectangle. The following facts enabled me to establish the geometry. It can be seen that the bottom centre of a third circular monument (henge N) is perfectly positioned at the bottom left hand corner of the double square. It has the same dimensions as the Knowth and Newgrange mounds and these three monuments all give the size of the square modules used to trace the plan. One can see how the East and West sides of the Knowth mound have been straightened to indicate the exact dimensions of the square. From the centre of Newgrange to the base of henge N we find a distance of 2 squares to the North and 14 squares to the West, the septuple square. The mauve line shows the angle of a 3-4-5 triangle, 12 squares at its base by 9 square high."[39]

Now if the Knowth to Newgrange angle is measured from due North we obtain 90°- 34.695° = 55.305° which is exactly three times the angle of a triple square, 18.435°. So the angle taken from North which aligns the Knowth and Newgrange monuments

---

38   See "Appendix 3 : The Moon's 18.6 year cycle", page 62

39   Howard Crowhurst, *Carnac, The Alignments*, p. 49, Epistemea, 2010

Figure 11. Knowth to Newgrange lunar geometry.

Figure 12. The Nebra sky disc orientated with the North at the top. The solid white lines show the major Moon positions and the dotted white lines show the minor Moon positions. The "Sun Ship" has become an arch.

at minor Moonrise is the equivalent of **six segments of the Nebra sky disc**.

As seen earlier, exact North is positioned on the disc at hole 20. In Figure 12, the disc has been rotated to position hole 20 at the top of the image. With this orientation, the Moon is now to the left of the Sun, showing its relative position at **minor** Southern moonrise.

 If we count six segments West from hole 20, we come to hole 14, inside the solstice arc, which is precisely marked by a gold dot. Counting six segment East from hole 20 brings us to hole 26, also marked by a gold dot. These angles have been traced from the centre using dotted white lines. They would consequently seem to place the Northern moonrise and moonset at its minor position, although this angle corresponds to the latitude at Newgrange and not at Nebra.

This would more probably indicate that the same geometrical method was used to trace the Nebra sky disc as that employed in Carnac and Newgrange.

However, astronomy software[40] shows that the angle of major southern moonrise in 1773BC at Nebra was precisely E53.13°S, which corresponds exactly to the greater angle of a 3-4-5 triangle. When measured from South, this becomes 36.87° which is 4 segments. Counting 4 holes West of hole 20 brings us to hole 16 and 4 holes East of hole 20 is hole 24. These angles have been traced from centre by white lines and it is most interesting to see that they position the extremities of the golden arc or "Sun ship". The two ends of the Moon crescent are also placed between major and minor Moon lines, showing its average position, half way between the two extremes.

____

40   J.Q. Jacobs and Victor Reijs, Archeogeodesy.xls.

THE SUN SHIP OR GOLDEN NUT ARCH?

The inner golden arc has been interpreted as a "Sun ship", well known in Egyptian mythology, chiefly because of its shape when seen at the bottom of the disc.

On the Nebra disc we see the ship together with the sun, helping the sun over the hills or through the night, through the underworld. I believe that the ship on the Nebra disc is the sun ship. [41]

However, when the disc is rotated the arc becomes an arch and could no longer be seen as a vessel. If compared to Egyptian art, it resembles Nut, the godess whose arched body represents the night sky (Figure 13).

The fact that the arc is positioned between the most Northern positions of the Moon puts it in a part of the sky beyond the reach of the zodiacal band, ie. the Sun, the Moon and the planets. This is the Northern sky where stars never rise nor set but turn eternally around the central point of the Heavens, marked today by Polaris, the Pole star. For this reason, these stars were known

---

[41] Dr. Flemming Kaul,(National Museum of Denmark), BBC documentary.

Figure 13. Representation of the Egyptian goddess Nut, whose arched body forms the sky.

as the "immortals" by the Greeks and the Egyptians. They were seen to part of another world, seperated from terrestrial life by the Styx, a raging river which had to be crossed to attain eternal life. The most well known constellations inside this eternal circle are the Plough (the Big Dipper or Great Bear) and Cassiopeia (shaped like a "W"). The fact that the inner circumference of this golden circle crosses the central point and does not go below the "horizon" would seem to be a clear indication of the "immortal" stars (Figure 14).

Was the size of this arc chosen arbitrarily? As it seems that it was added after the other elements on the disc, one could suppose that it was positioned so that it would not cover the pre-existing gold dots. However, having seen the care which had been taken in the disc's design and fabrication, it would be fairer to assume that this golden arc, even if it was added later, was sized and placed in some meaningful way. This must have been an extremely precious object, on a par with the Crown Jewels or the Mona Lisa, and would not have been available to "amateurs" (unless they stole it).

Figure 14 shows a study of the arc's characteristics. Two yellow circles indicate the inner and outer circumferences of the arc and show that it is perfectly circular. The outer diameter is very close to half the size of the Nebra sky disc (16.1cm), possibly indicating the Northern sky as half of the heavens, and the inner diameter has been measured to be 13.95cm. The first thing one notices is the fact that the arc is inclined compared to the horizontal and vertical structure previously identified. The blue line, joining the two solstice limits on the disc's circumference, exactly touches the inner left end of the arc but is much higher than the right end. A line drawn between the two ends is tilted by 7.5° (one 48th part of a circle) from the horizontal blue line. The central point of the arc is at hole 20, which is also the central point of the disc's vertical structure, clearly showing a link between the two "systems". This may well symbolise the fact that the Northern sky is indeed inclined (unless one is standing exactly at the North

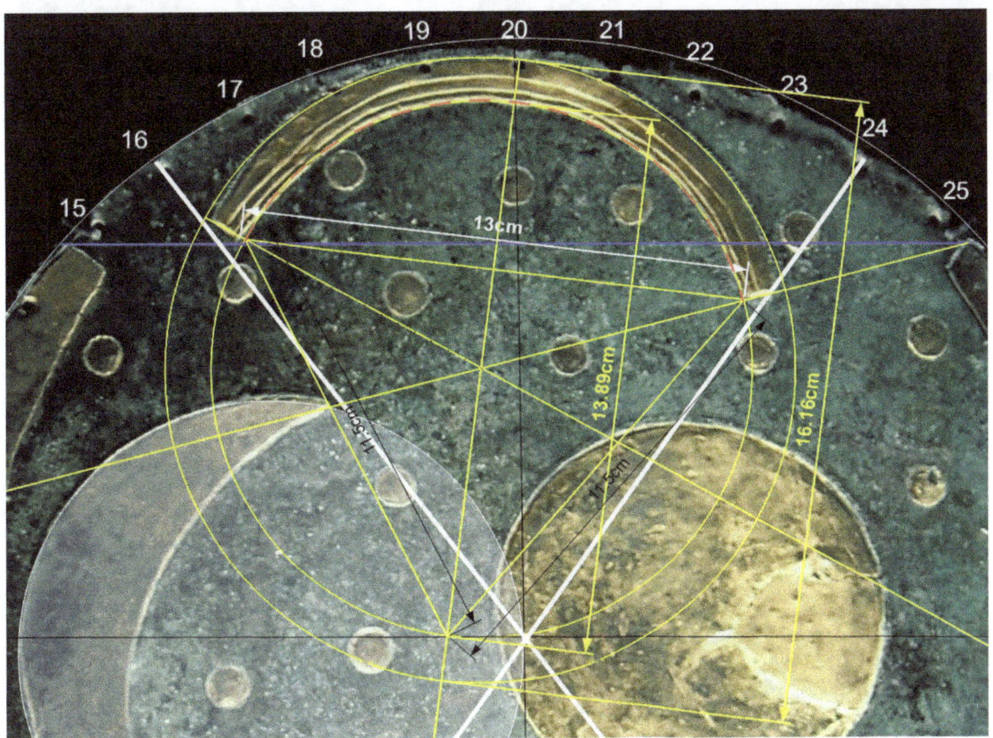

Figure 14. The geometry of the golden arch. The portion of the circle which has been drawn is in a golden ratio to the invisible portion.

Pole). The arc's two extremities have been cut at right angles to the circles, which means that they indicate the centre.

A line drawn from the upper tip of the crescent Moon, through this centre and then through the right end of the arc joins the blue line at the exact solstice point on the right, so although the arc is inclined, its extremities are both linked to the solstice positions. If this is intentional, which would seem obvious, then the size and position of the arc was determined so that this would be the case. The centre of the arc is the position of the Pole Star. The point on the projected perimeter opposite hole 20 is equidistant (11.5cm) from the two ends of the arc. Two yellow lines on the diagram indicate this symmetry.

Now the chord of the inner arc can be seen to have exactly the same length as the diameter of the Moon circle, 13cm, which confirms the intentional nature of these dimensions. But what proportion of its own circle does the arc describe (indicated on the diagram by the dotted red and yellow line)? This is shown by the angle at the centre of the circle which equals 137.5°.[42] The part of the circle which was not represented is equal to 222.5° (360° - 137.5°). So the proportion of the invisible part to the visible part is 222.5° divided by 137.5° which gives the result 1.618.

This proportion is known as the divine proportion or golden ratio and was used extensively in ancient Egyptian architecture, art and even calendar construction. It is naturally present in nature and determines the way things grow. One example of this is phyllotaxis, the spiral arrangement of leaves on the stem of a plant.

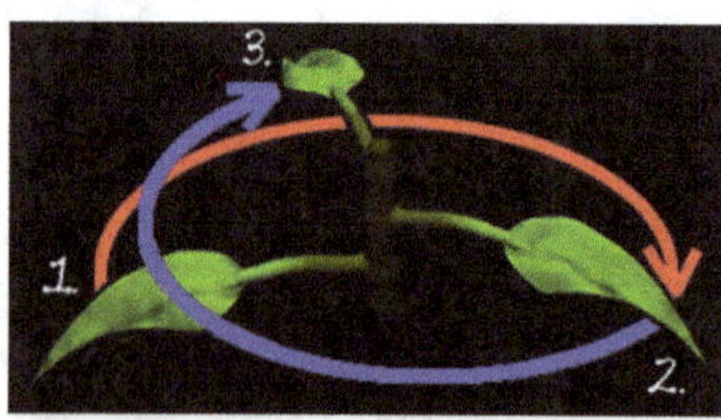

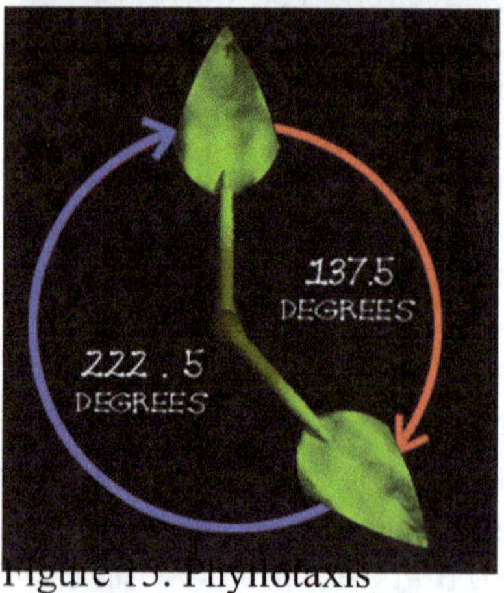

Figure 15. Phyllotaxis

In an overwhelming number of plants, a given branch or leaf will grow out of the stem approximately 137.5 degrees around the stem relative to the prior branch. In other words, after a branch grows out of the plant, the plant grows up some amount and then sends out another branch rotated 137.5 degrees relative to the direction at which the first branch grew out.[43]

It is beyond the scope of this present work to develop more on the fascinating subject of the divine proportion[44], expressed

42   137.5°= 2xsin⁻¹(13/13.95) = 4xsin⁻¹(13/23)

42        $137.5° = 2 \times \sin^{-1}(13/13.95) = 4 \times \sin^{-1}(13/23)$

43   Text and diagram from http://www.natures-word.com/sacred-geometry/phi-the-golden-proportion/phi-the-golden-proportion-in-nature

44   If the angle of 225.5° is divided by 24, the number of hours in a day, the result is 9.27° which is close to the division of a circle by 39 (9.23°)

50

several millennia earlier in a totally different way in the Carnac alignments.[45] The presence of this golden ratio, inscribed on a golden arc on the Nebra sky disc, is further evidence of its designer's high level of knowledge.

It has been suggested[46] that the arc angle corresponds to the part of the year when Pleiades is visible on the western horizon. As this has been calculated to last 151 days (see page 73), and despite the fact that it is a good idea, the hypothesis is erroneous by over 10°.

45   Howard Crowhurst, Carnac, The Alignments, p. 28-33, Epistemea, 2010.

46   Goseck and Nebra, two cosmic representations, http://users.skynet.be/lotus/tools/nebra0-en.htm, unknown author.

MERCURY AND OTHER PLANETARY CYCLES

Another planet whose synodic cycle is close to a multiple of 39 is Mercury. As it is close to the Sun, its moves much faster and returns to the same position in relation to the Sun every 115.88 (116) days. Now 39 times 3 is 117 days. So if a Mercury peg is moved every 3 days, after one lap the planet has returned to its initial position plus 1 day. Mercury's name in Greek was Hermes Trismegistus. According to Wikipedia :

"The origin of the description Trismegistus or "thrice great" is unclear."

Perhaps the title comes from ancient calendar practices and the link between the Mercury cycle and the 3-day period, one 39th of its cycle.

After 5 Mercury laps, or 585 days, Mercury will be in the same position as it was in relation to Venus and after 20 laps or 2340 days, it will be in the same relationship with Venus and Mars, as seen above, but with a 20 day difference (or one Sun peg). So 20 Mercury cycles (plus 20 days) equals 4 Venus cycles (plus 4 days) equals 3 Mars cycles and the best common denominator of these incredible coinciding cycles is 39, the number of holes equally spaced around the Nebra sky disc.

There may have been other cycles which could have been followed on the disc. Jupiter, with its synodic period of 399 days, would be a good candidate. Its cycle could be followed by moving a Jupiter peg one hole every 10 days then pushing it back a hole after one lap and one hole. As we have seen, 39 is also 3 times 13, so if a second Moon peg was moved forward one hole every 9 days, 3 holes would mark one lunar orbit of 27 days, and one lap would indicate the 13 moon orbits (sidereal revolutions) in a year. 27 days is also 20 days plus 7 days, a score plus a week, linking the disc's two major cycles.

The brightest asteroid in the sky, known as 4 Vesta, is visible to the naked eye when it is in opposition to the Sun. It has an orbital period of 1325.85 days which is 34 times 39 days. However, without the instruction manual, it is difficult to know exactly how many cycles could be followed.

The different cycles related to the number 39 are summarised in the following table.

	Period (days)	Type	Multiple of 39
Sun (axial rotation)	27.3	Synodic	7/10
Mercury	115.9	Synodic	3
Venus	583.9	Synodic	15
Sun	365.26		28/3
Moon	27.3	Sidereal	7/10
Mars	779.9	Synodic	20
4Vesta	1325.8	Sidereal	34
Gestation	273		7

Once the Nebra sky disc has been calibrated, with the different solar, lunar and planetary pegs in their correct positions, it then becomes possible to use the device for prediction. By advancing each peg according to its relative speed, the future movement of the heavenly bodies is revealed and conjunctions can be determined. It then becomes possible to organise a religious calendar and determine feast days and fast days accordingly. One can begin to understand the importance of this object and the great care which had been taken in making it.

MEASUREMENTS AND THE METRE LENGTH

A subject which no researcher has dared approach concerns the dimensions of the disc and the possibility that its measurements could give us a clue to the measurement system in use at the time of its fabrication. As the author has not been able to measure the disc personally, all suggestions made hereafter are subject to caution and would need to be checked on the object itself. It is also hoped that the photograph used has not been too distorted. On the image, the disc is not a perfect circle, being slightly flattened "at the poles". This being said, its horizontal diameter is given as being 32cm and this measurement has been used a basis of the study. The relevant diagrams are Figure 9, page 39 and Figure 14, page 49.

It has been suggested in this present work that the disc was traced using the triple square and the 3-4-5 triangle,[47] the 4 side of the triangle giving the radius of the disc. As the radius is 16cm, which is divisible by 4, this would indicate a unit of 4cm for the 3-4-5 triangle. The next interesting fact is that the diameter of the golden Sun disc is 10cm and that the circumference of the whole disc is 10 times that (32 times Pi), making 1 metre. In other words, if the disc was rolled on the ground, each rotation would measure out one metre. Now one inch, a very ancient measurement, is 2.54cm and 39 times this makes 99.06cm so the average distance between the centre of the holes, which are not quite on the edge of the disc, is one inch. The golden dots, with an example shown as 1.27cm, appear to measure half an inch. The Moon has a diameter of 13cm and this could be seen as strange since 13 is a number with strong lunar associations. This is also the chord of the Nut arch. We have already seen an incredible coincidence in the use of centimeters for the 7 by 8 solstice rectangle which measures 21cm by 24cm on the disc, a unit of 3cm.

It is important to point out here that the number relating the centimetre to the inch, 2.54, is a luni-solar number as there are 254 moon orbits in 19 years, one Metonic cycle.[48]

---

47   See "Appendix 2 : Dividing a circle into 39 equal parts", page 61
48   See "", page 66

These different findings can either be seen to be coincidences or indicate that the people who created this disc had much greater knowledge and science than has ever been imagined, on a par with the pyramid builders who had preceded them.

In conclusion, the numbers, geometry and symbolism on the Nebra sky disc infer that the knowledge of astronomy that it represents could have been inherited from the megalith builders several millenia earlier. The proximity of the Goseck circle, which marks the solstices and has been dated to -5000, could be seen as tangible evidence in support of this hypothesis. The division of the circle into 39 parts shows the understanding of some kind of cohesion underlying the solar system which we no longer recognise but which was the foundation of the later Mayan calendar system and the Pythagorean school of Ancient Greece.

Postface : The 58-hole Game

After having published my discoveries concerning the Nebra Sky Disc, I was contacted by a reader who informed me of the existence of an ancient "game" board, which has been named "the 58-hole game". Examples of it have been found in Egypt and in several Middle East countries and can be seen in Figure 17.

Despite its name, this board actually has 59 holes (one of the holes is bigger than all the others), 20 in the middle and 39 around the edge ! It would seem to be an incredible confirmation of the ancient method of computing planetary movement, combining the 20-base vigesimal system with the 13 base system.

Moreover, by completing a whole "lap" around the board, going from hole n°1 to hole n°30 and then coming back to hole n°1 on the other side (Figure 16), moving a peg one hole per day, two full moon cycles of 29.5 days are counted, a total of 59 days.

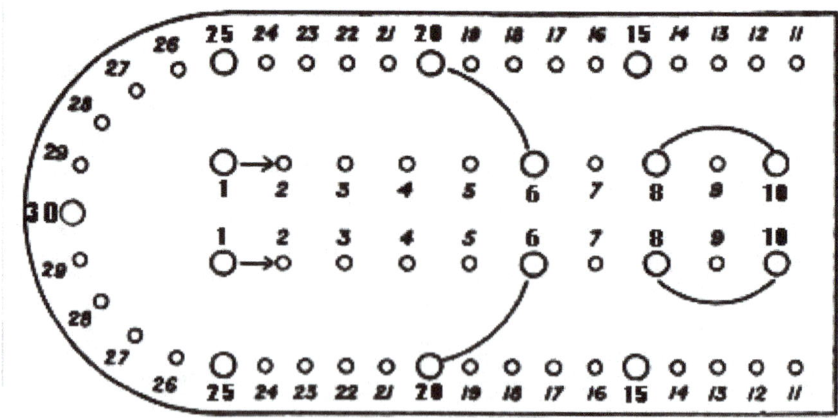

Figure 16. The organisation of the 59 holes on the board

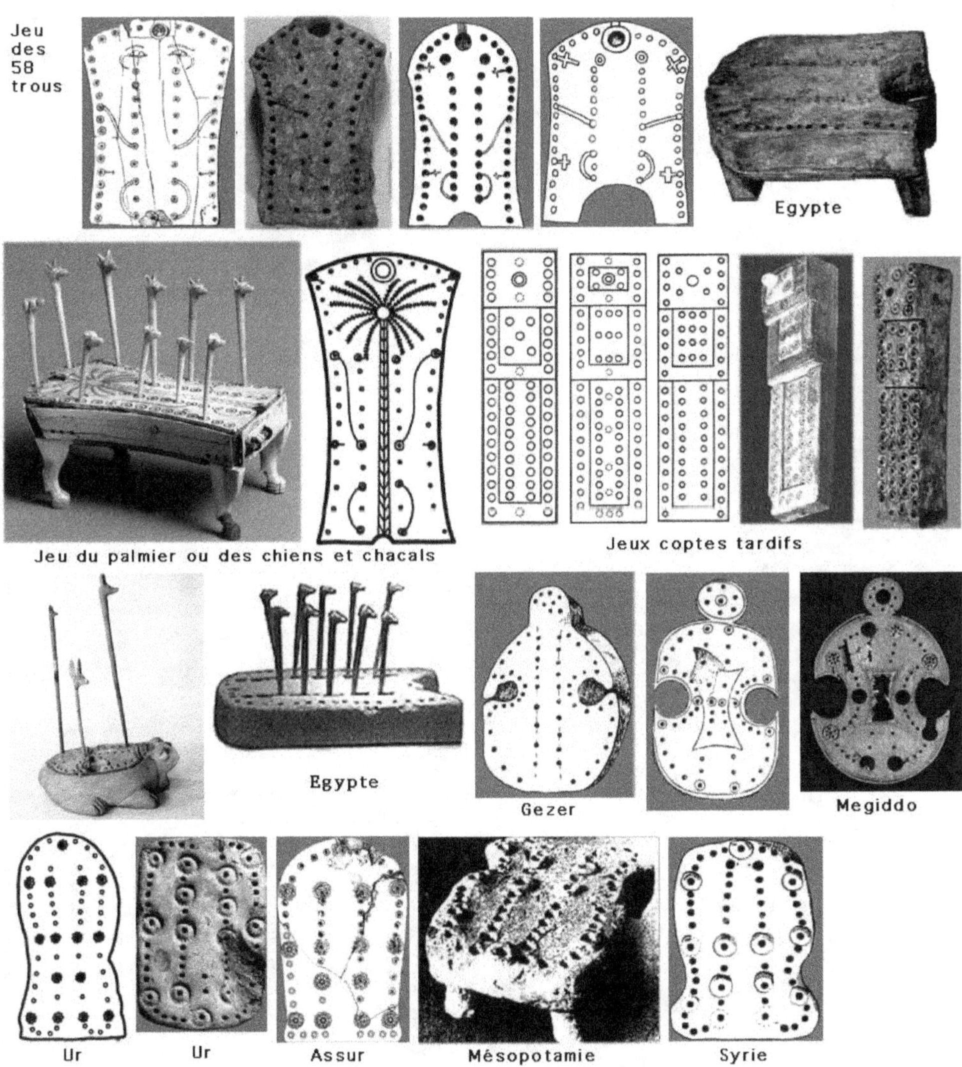

Figure 17. Different examples of the "58-hole Game".

APPENDICES

APPENDIX 1 : SOLSTICES

Everybody who lives in the Northern hemisphere above the tropics knows that days last longer in summer than in winter. In summer, the Sun rises higher in the sky. It also rises and sets further to the North. If one goes far enough North, one can come to a point, on what is called the Arctic circle, where the Sun no longer even sets at midsummer and can be seen sitting on the horizon due North at midnight.

In winter, the opposite is true. The Sun rises and sets towards the South and is lower in the sky at midday. North of the Arctic circle, it is so low that it never actually rises above the horizon and the land is cloaked in permanent dark.

These seasonal variations are thus accentuated the further North one goes, but on all European latitudes, they greatly affect everyday life. The extreme Northern and Southern positions of the sunrises and sunsets occur at moments called solstices. Summer solstice, or midsummer, 21st June, corresponds to the longest day of the year, when the Sun rises and sets furthest North and winter solstice, midwinter, 21st December, is the shortest day of the year when the Sun rises and sets furthest South. The word solstice comes from Latin and means "Sun stops" or "Sun stations", because at these extreme positions, the Sun can be seen to rise or to set in the same place for several days.

Half way between these two extremes is the East-West axis, where the Sun rises and sets at equinoxes. These points mark the beginning of Spring and Autumn, when days and nights last the same time.

So according to latitude, the distance between the extreme positions changes. The further North one goes, the greater the angle between the two solstice positions.

The angular value of 82° accurately corresponds to the difference between winter and summer solstice positions at the latitude of Nebra. This can be divided into two equal parts, each of 41°, above and below the East-West Line. So the summer solstice sunrise position can be expressed as 41° North of East or E41°N and the winter solstice position as 41° South of East or E41°S.

Appendix 2 : Dividing a circle into 39 equal parts

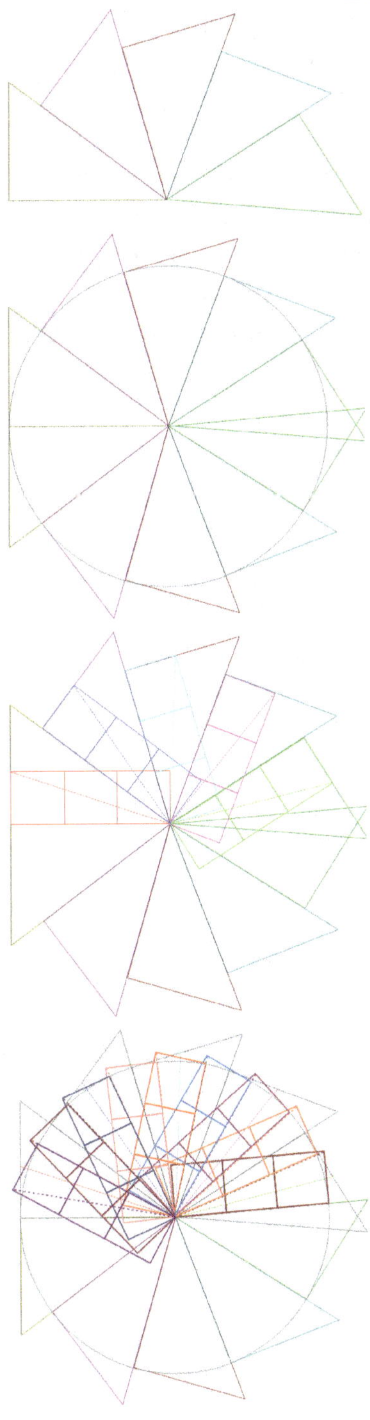

(1) A 3-4-5 triangle, with its 4 side horizontal, is rotated four times around its acute angle in a clockwise direction.

(2) This process is mirrored around the horizontal axis. A circle is drawn which goes through all the right angles. Opposite the initial base line, the two intersecting triangles create the first 39th segment.

(3) A triple square, a one by three rectangle (in red), whose long side is equal to the radius of the circle, is placed on its horizontal radius and its diagonal is traced (dotted line). This exactly bisects the angle of the 3-4-5 triangle. A second triple square (in blue) with its diagonal is placed along the base of the next 3-4-5 triangle clockwise. This process is repeated for the five upper triangles.

(4) A triple square (in dark red) is placed along the **hypotenuse** of the next 3-4-5 triangle, the fifth of the lower section, and its diagonal is drawn from the centre of the circle. The base of another triple square (in orange) is placed along this diagonal, coming back anti-clockwise around the circle, and again its diagonal is traced. This process is repeated 9 times until one returns to the initial base line of the first 3-4-5 triangle. A little more than half the circle, 20 segments out of 39, has been traced. The remaining 19 segments can be traced in the same way for the lower part of the circle.

APPENDIX 3 : THE MOON'S 18.6 YEAR CYCLE

This cycle, which is due to the precession of the Moon's nodes, can be measured by observing the varying amplitude of the Moon's rising and setting. It has two extreme positions; major, which corresponds to the Moon's most Northerly and Southerly rising and setting over this 18.6 year period and minor which corresponds to its shortest amplitude. The major Moon rises and sets outside the Sun's solstice arc whilst the minor Moon rises and sets well inside of it. Professor Alexander Thom put forward conclusive evidence that the architects of megalithic monuments had used these extreme lunar positions.

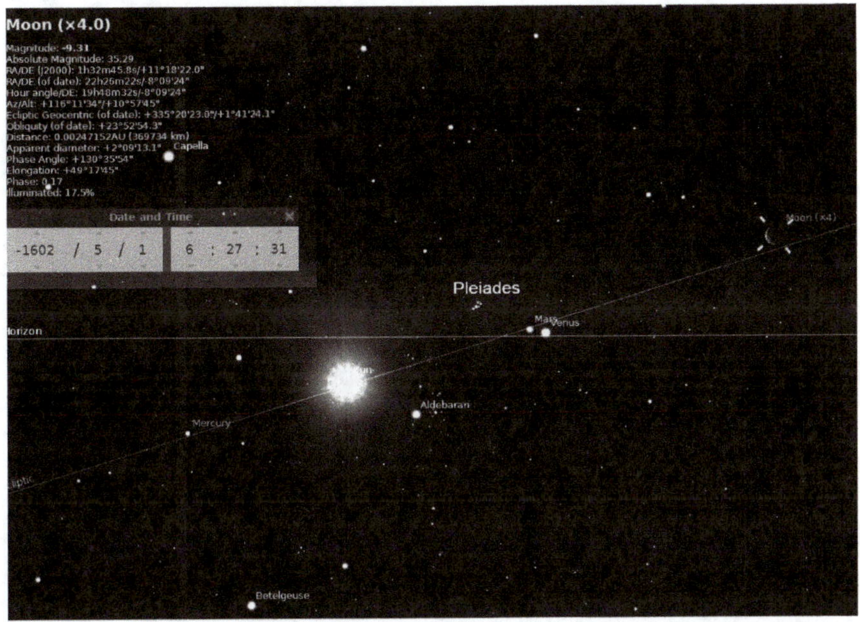

As it is clear that the Pleiades are represented on the Nebra sky disk, between the sun and a crescent moon, it becomes possible that it shows a particular configuration of the Sun, Pleiades and the crescent moon, which would take place every 19 years on the 1st May, when the Sun and the Moon would be in exactly the same positions. On this particular occasion, three gold dots can be seen between the Sun and the Moon, and these would have to be on the ecliptic. The lower of the three could be the star Aldebaran, the Bull's Eye, the brightest star in the Taurus constellation. The two gold dots above, however, would have to be planets, as would the dot on the other side of the Sun, because there are no other bright stars near the ecliptic in this stellar region.

A possible example is shown in the figure opposite left (courtesy of *Stellarium*). It is a view of the night sky dated from the 1st May 1602BC and valid for the latitude at Nebra. We see the Sun on the left, Pleiades in the centre and the crescent moon on the right. Between the two, Mars and Venus are in conjunction just above the horizon.

The sun is below the horizon and so would not have been visible. Nor would Mercury, which is just behind the Sun. This means that the whole scene on the disk would have been reconstructed through a mental representation. A crescent moon would never be orientated in the way shown on the disk because the sun illuminates the side of the Moon which is facing it. Also, Aldebaran, Mars, Venus and the lower stars in the Pleiades could never be seen as shown on the disk because the Moon is nearer the Earth than them and its non-illuminated part would hide all the celestial objects behind it. This shows clearly that the image is a representation. How could this have been done ?

The exact position of the Pleiades can be easily calculated thanks to an extraordinary coincidence. The star group's position in the zodiac is almost exactly 90° (7 minutes difference) around the belt from one of the brightest stars in the sky, Regulus in the constellation of Leo, the Lion. In Arabic, this star is called Kalb Al Asad , the Lion's Heart. It is the closest bright star to the ecliptic, the path of the Sun, which means it is regularly occulted by the moon, and sometimes the planets, making it an invaluable companion for

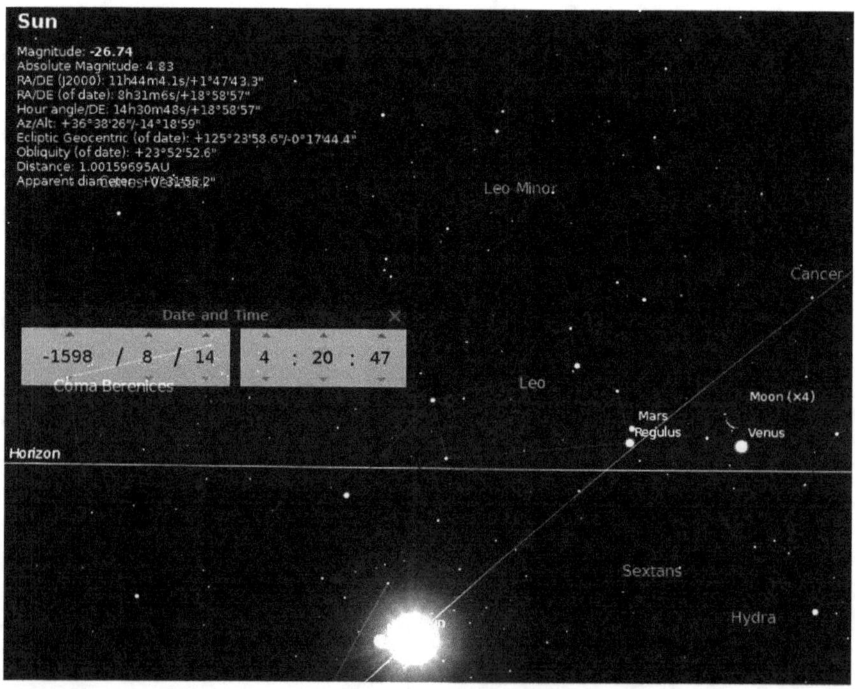

the precise calculations of solar, lunar and planetary cycles. It also means that when Regulus cumulates, at its highest point in the sky due South, it is at the exact co-latitude angle (90° minus latitude) allowing accurate geodesic calculations. Finally, when Regulus cumulates just before dawn, the Pleiades are rising in the East. As Regulus is brighter than the Pleiades, with a magnitude of +1.35, and at its highest point, it is perfectly visible and so "announces" the appearance of the Pleiades with great precision.

The sky chart below represents the sky on the 14th August -1598, 40 days before Autumn equinoxe and 6.4 years after 1st May -1602, the configuration shown in the first sky chart. Venus and Mars are in conjunction again, just near a crescent Moon, but this time at the heliacal rising of Regulus.

APPENDIX 5 : THE WELSH CONNECTION?

The Isle of Mona, Moon in Old English, now known as Anglesey, on the North-West coast of Wales, was an ancient Druidic stronghold before Roman occupation. On either side of the shield on the coat of arms are two animals, a bull on the left and a lion on the right. They are both white, which besides being a druidic symbol, could also mean they are a source of light. Could this be a representation of the constellations of Taurus and Leo and their celestial signatures, the Pleiades and Regulus?

APPENDIX 6 : THE HYPERBOREANS

Diodorus of Sicily talks about the existence of a people called the Hyperboreans, saying that they had been in contact with the Greeks from very ancient times.

In the regions beyond the land of the Celts there lies in the ocean an island no smaller than Sicily. This island, the account continues, is situated in the north and is inhabited by the Hyperboreans, who are called by that name because their home is beyond the point whence the north wind (Boreas) blows; and the island is both fertile and productive of every crop, and has an unusually temperate climate...There is also on the island both a magnificent sacred precinct of Apollo and a notable temple which is adorned with many votive offerings and is spherical in shape...The Hyperboreans also have a language, we are informed, which is peculiar to them, and are most friendly disposed towards the Greeks, and especially towards the Athenians and the Delians, who have inherited this good-will from most ancient times. The myth also relates that certain Greeks visited the Hyperboreans and left behind them there costly votive offerings bearing inscriptions in Greek letters. And in the same way Abaris, a Hyperborean, came to Greece in ancient times and renewed the good-will and kinship of his people to the Delians. They say also that the moon, as viewed from this island, appears to be but a little distance from the earth and to have upon it prominences, like those of the earth, which are visible to the eye. The account is also given that the god visits the island every nineteen years, the period in which the return of the stars to the same place in the heavens is accomplished; and for this reason the nineteen-year period is called by the Greeks the "year of Meton." At the time of this appearance of the god he both plays on the cithara and dances continuously the night through from the vernal equinox until the rising of the Pleiades, expressing in this manner his delight in his successes.[1]

The mention of the 19 year Metonic cycle which was purportedly introduced in Athens in 432 B.C. and was designed to reconcile

---

1   Diodorus Siculus, Book II, 47-48

the lunar and the solar year, is noteworthy. Diodorus speaks of votive offerings given by the Greeks to the Hyperboreans. *Votivus* in Latin means promised by a vow, implying that the Greeks had agreed to offer costly gifts to the Hyperboreans in exchange for some service. He then immediately goes on to talk about the Moon. The text leaves one to deduce, then, that the Hyperboreans were aware of the metonic cycle before Meton himself and one may even infer that the Greeks learnt of it from them. We in fact know very little about Meton as none of his works have survived. In the play by Aristophanes, The Birds, he appears on stage with surveying equipment and is described as a geometer, not an astronomer.

Diodorus actually says that the reason the Greeks call the cycle "year of Meton" is because the stars return to their same place in the heavens. He does not say they named it after someone. Now *meto* in Latin, Diodorus' language, means "I measure". This word originates from the PIE root *met*, measure, from which derives the ancient Greek word *metron* and gives us the word metre. *Metis* in Greek means cunning and skilful but also a craft. *Metor* in Latin means to mark out a field, the typical work of a geometer. Perhaps one may conclude that there is a possibility that the cycle of Meton means the measured cycle and that in order to be precisely measured, a geometer, or Earth measurer, *meton* in Greek, was needed to mark out the exact positions of sun and moon rises. The metonic cycle of 19 years corresponds to a high degree to the same Moon phase and Moon position in the stars on the same day of the year.

The text finishes with a description of a festival which takes place between vernal equinox and the rising of Pleiades. At the time of the Nebra sky disc, this festival would have lasted 40 days, between the 21st March and May 1st.[2] The night preceding May 1st was a major festival in Northern European countries and is known as Walpurgis Night.

---

2   See Appendix 7.

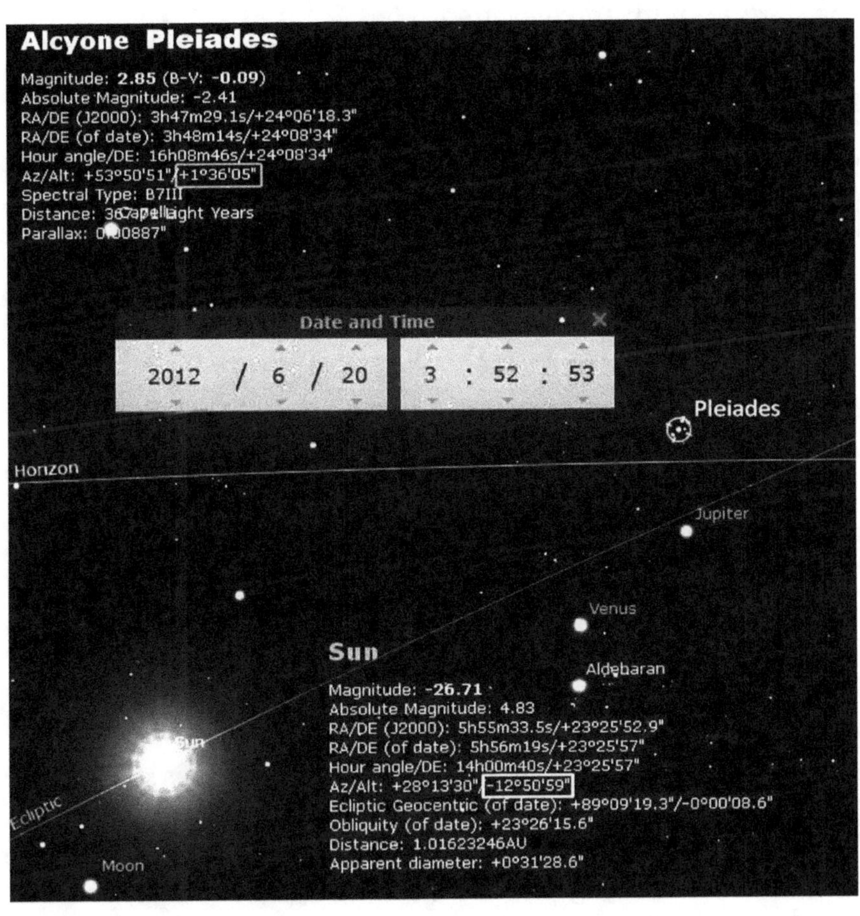

Alcyone Pleiades
Magnitude: 2.85 (B-V: -0.09)
Absolute Magnitude: -2.41
RA/DE (J2000): 3h47m29.1s/+24°06'18.3"
RA/DE (of date): 3h48m14s/+24°08'34"
Hour angle/DE: 16h08m46s/+24°08'34"
Az/Alt: +53°50'51"/+1°36'05"
Spectral Type: B7III
Distance: 367 light Years
Parallax: 0.00887"
Date and Time
2012 / 6 / 20   3 : 52 : 53
Pleiades
Horizon
Jupiter
Venus
Sun
Aldebaran
Magnitude: -26.71
Absolute Magnitude: 4.83
RA/DE (J2000): 5h55m33.5s/+23°25'52.9"
RA/DE (of date): 5h56m19s/+23°25'57"
Hour angle/DE: 14h00m40s/+23°25'57"
Az/Alt: +28°13'30"/-12°50'59"
Ecliptic Geocentric (of date): +89°09'19.3"/-0°00'08.6"
Obliquity (of date): +23°26'15.6"
Distance: 1.01623246AU
Apparent diameter: +0°31'28.6"
Sun
Ecliptic
Moon

APPENDIX 7 : CALCULATING THE HELIACAL RISING AND SETTING OF PLEIADES

The factors which determine the visibility of a star on the horizon are the following:
The star's brightness
The star's azimuth and elevation
The angular distance of the Sun beneath the horizon
Atmospheric conditions
Light pollution
In the case of observations in the past, we can eliminate the last two factors from our calculations.
The first two factors compose what is known as the arcus visionis. This can be calculated using tables created by A.F. Aveni[3]. Here is an extract of the parameters used by Aveni (1972) in constructing arcus visionis, adapted from Lockyer (1894).

Star and Sun on same horizon (RSR, SSS)

Magnitude	Solar Altitude	Altitude of star
1	-11°	+1°
2	-14°	+2°

Star and Sun on opposite horizon (SSR, RSS)

Magnitude	Solar Altitude	Altitude of star
1	-7°	+1°
2	-8.5°	+2°

Four types of heliacal star observations are possible, as seen above.
Two heliacal star risings:
1) RSR - (Rising-Sun Rise) when the star appears rising above the horizon to the East slightly before sunrise. (Star close to Sun)
2) SSR (Setting-Sun Rise) when the star appears rising above the horizon to the East just after sunset. (Star opposed to Sun)
Two heliacal star settings:
3) RSS (Rising-Sun Set) when the star appears setting above the horizon to the West slightly before sunrise. (Star opposed to Sun)

---

3   Aveni, A.F. 1972, Astronomical tables intended for the use of astro-archaeological studies, american antiquity xxxvii, 531-40

4) SSS (Setting-Sun Set) when the star appears setting above the horizon to the West, just after sunset. (Star close to Sun)

The star cluster of the Pleiades contains seven stars with the following magnitudes

Star	Magnitude
Alcyone	2.87
Atlas	3.62
Electra	3.7
Maia	3.86
Merope	4.17
Taggeta	4.29
Pleione	5.09

To calculate the magnitude of the cluster, the following formula is used:

$$e = \sum_{i=1}^{7} 10^{-m_i/2.5} \quad \text{then} \quad m = -2.5\log_{10} e = 1.6$$

As the Pleides cluster has a magnitude of 1.6, this enables us to determine the following values for its "close to Sun" rise and set.

Solar altitude = -12.8° = 12°48' (below horizon)

Star Altitude = 1.6° = 1°36' (above horizon)

Total angular distance= 14.4°

Similar calculations can enable us to determine the "opposed to Sun" rise and set. The angular distance of the Sun below the horizon is much smaller as it is opposite the position of the star and so its brightness is greatly diminished. The figures for the Pleiades are as follows:

Solar altitude = -7.9° = 7°54' (below horizon)

Star Altitude = 1.6° = 1°36' (above horizon)

Total angular distance= 9.5°

The most well known heliacal observation in history is 1) RSR, known as the return of the star after its occultation by the Sun. The star becomes visible for several minutes before sunrise, which puts an end to it's period of invisibilty. The Sirius RSR determined the beginning of the year in Ancient Egypt. In 2012 at the latitude of Stonehenge, this event occurred for the Pleiades on the 20th June at

3:52:53 (see image above). On the 19th June, visibility was not possible. On the 21st June, it was easier. Every day after that, the cluster remained visible for longer, rising earlier by 2 hours per month, (24 hours in 12 months) until we came to the event 2)SSR when it rose just after sunset. This happened on the 29th october 2012 at 18:16. The next day, the Pleiades rise was no longer visible because dusk was too bright.

This corresponds to the beginning of a period when the Pleiades cannot be seen to rise or set because both events happen during the daylight hours. Since the star cluster is opposite the Sun, it rises when the Sun sets and sets when the Sun rises. However, it is perfectly visible during the night and this is the best time to see it clearly. This period is shorter than the occultation period (see below) as the angular distance of invisibility is smaller, as explained above. The setting of the Pleiades in the West, 3)RSS, will be possible on December 3rd 2012 at 7:36, making a total of 34 days and 35 nights of invisibility on the horizon. From then on, the setting will be visible every day, taking place 2 hours earlier each month until we get to 4)SSS on May 3rd 2013 at 22:28, five months afterwards. This is the last time the star will be visible before its occultation by the Sun. It is the start of the star's period of invisibility.

In 2012, the last evening the Pleiades were visible to the west after sunset was 3rd May at 22:41:30. Again, on the 4th May, they were no longer visible. So, even in perfect weather conditions at the latitude of Stonehenge, the Pleiades remained invisible between the evening of 3rd May until the morning of the 20th June, a total of 47 days and 48 nights. The following table resumes these events.

Pleiades Event	Present Date	Time after precedent	Description
1) RSR	20/06/2012	48 days	Star's return, end of occultation
2) SSR	29/10/2012	131 days	Disappearance of star on horizon (East)
3)RSS	3/12/2012	35 days	Reappearance of star on horizon (West)
4)SSS	3/05/2012	151 days	Star disappears. Start of occultation

When Professor Harald Meller, director of the National Prehistory Museum of Saxony-Anhalt in Halle says: *"The Pleiades, which are usually depicted as six or seven smaller stars, are especially visible in the western sky between 17 October and 10 March"*, he is not talking about the present day but about some time in the past.

When Hesiod remarked, around -700, that the Pleiades were invisible for forty days and forty nights in Greece, he can be seen to have been correct, as at a lower latitude the ecliptic is more vertically inclined and so the angles are reduced, making occultation shorter.

At Stonehenge, then, a Pleiades rising is visible about 24 days after the solar conjunction.

Because of the precession of the equinoxes, the heliacal observation dates slide back slowly in time. The constellations retrograde by 50.290966 arc-seconds per year or 1 day every 70.556 years. At Hesiod's time, 2712 years ago, they took place 38 days earlier than at present, so RSR was on May 13th -700. This is the time he suggests to start harvesting (page 38). His date for ploughing would be either 26th October or 26th March, depending on whether he was suggesting pre-winter ploughing or not.

The RSR of the Pleiades took place on the 1st May, 50 days earlier than at present, 3528 years ago (50 x 70.556) or in -1516. At the latitude of Nebra and Stonehenge, this would be far too early to begin the harvest. The RSS to SSS period in -1516 started 50 days before December 3rd, which is October 14th. This must correspond to the dates spoken of by Harald Meller above.

RSR took place at vernal equinox on the 21st March, another 40 days earlier than at Nebra, 2822 years before that, in -4338. Consequently, this is the earliest possible date for the Hyperborean festival (see page 66). When Diodorus says *"he both plays on the cithara and dances continuously the night through from the vernal equinox until the rising of the Pleiades"*, were he to mean that the heliacal rising of Pleiades took place the next morning, 4300BC would be the date of the origin of this story. This just happens to correspond to the beginning of the megalithic period in Brittany. This astronomical event would then happen 1 day later every 70.6 years, which would mean that by -1600, the date of the Nebra sky disk, as we have seen, the feasting would have lasted 40 days until **May 1st**.

Table of Figures

www.ingramcontent.com/pod-product-compliance
Lightning Source LLC
LaVergne TN
LVHW020952200726
843508LV00004B/1407

9 782379 000331